1986 Supplement

Publications of the International Agricultural Research and Development Centers

1986 exhibition
at the Frankfurt Book Fair
sponsored by:

Deutsche Gesellschaft für
Technische Zusammenarbeit (GTZ) Gmb H
(German Agency for Technical Cooperation) (GTZ)

Consultative Group on
International Agricultural Research (CGIAR)

International Rice Research Institute (IRRI)

1986
The International Rice Research Institute
Los Baños , Laguna , Philippines
P.O. Box 933 , Manila , Philippines

The International Rice Research Institute (IRRI) was established in 1960 by the Ford and Rockfeller Foundations with the help and approval of the Government of the Philippines. Today IRRI is one of the 13 nonprofit international research and training centers supported by the Consultative Group for International Agricultural Research (CGIAR). The CGIAR is sponsored by the Food and Agriculture Organization (FAO) of the United Nations, the International Bank for Reconstruction and Development (World Bank), and the United Nations Development Programme (UNDP). The CGIAR consists of 50 donor countries, international and regional organizations, and private foundations.

IRRI receives support, through the CGIAR, from a number of donors including: the Asian Development Bank, the European Economic Community, the Ford Foundation, the International Development Research Centre, the International Fund for Agricultural Development, the OPEC Special Fund, the Rockefeller Foundation, the United Nations Development Programme, the World Bank, and the international aid agencies of the following governments: Australia, Canada, China, Denmark, France, Federal Republic of Germany, India, Italy, Japan, Mexico, Netherlands, New Zealand, Norway, Philippines, Saudi Arabia, Spain, Sweden, Switzerland, United Kingdom, and United States.

The responsibility for this publication rests with the International Rice Research Institute.

ISBN 971-104-145-6

Introduction

The International Agricultural Research Centers (IARCs) focus modern agricultural research on the crops and livestock that provide 75% of the food for developing nations. The Centers are major publishers of books, periodicals, slide sets, films, and other educational materials on agricultural science and technology for developing nations.

In 1985 the International Rice Research Institute published *Publications of the International Agricultural Research and Development Centers*, a 691-page catalog of titles published by 13 Centers supported by the Consultative Group on International Agricultural Research (CGIAR); 8 other IARCs; the Board on Science and Technology for International Development (BOSTID) of the U.S. National Academy of Sciences; and the German Agency for Technical Cooperation (GTZ). The catalog includes 162 pages of in-depth index to help the reader locate all publication in certain fields (i.e. cytogenetics, insect resistance, maize). Ordering instructions for the 1985 catalog are on page 167 of this Supplement.

The 1986 Supplement includes only new titles that are not in the larger 1985 catalog. The two catalogs are the only compilations of the major Center publications and, collectively, are probably the largest catalog of titles on Third World agriculture in existence.

We have chosen to release the Supplement at the exhibit of Publications of the International Agricultural Research and Development Centers at the 1986 Frankfurt Book Fair. This is the fourth such exhibit sponsored by the CGIAR, GTZ, and IRRI.

We hope that the catalogs will bring us closer to the worldwide community of publishers and distributors in agricultural science. A wider exchange of knowledge on international agricultural research and production will help us achieve our ultimate objective — to make hunger and poverty a problem of the past.

Dr. K.J. Lampe
Head, Department of Agriculture
Health and Rural Development
Deutsche Gesellschaft für Technische
Zusammenarbeit (GTZ) GmbH

Dr. M.S. Swaminathan
Director General
International Rice
Research Institute

Contents

Part II

Non CGIAR-Supported Centers

Part III

CGIAR

Consultative Group on International Agricultural Research

1818 H St., N.W. Washington, D.C. 20433 USA
Cable Address: INTBAFRAD - Telex Worldbank 440098
Telephone: 202-477-5347

What is the CGIAR?
What do the international centers do?

The Consultative Group on International Agricultural Research (CGIAR) is a relatively new and exciting strategy to increase global food production. Established in 1971, the CGIAR is an association of 48 countries, including the Federal Republic of Germany, international and regional organizations, and private foundations supporting 13 international agricultural research institutes. The purpose of the Consultative Group is to bring the resources of modern biological and socioeconomic research to bear on the problems of improving agricultural productivity in the tropics and subtropics where most of the developing nations lie. The research and training programs undertaken by the centers and sponsored by the Group seek to arm the developing countries with superior varieties of essential crops and improved farming systems for the production of food plants and animals. The food crops and livestock on which the centers focus provide 75% of the food consumed in the developing countries. Located in developing countries, the international centers work together with scientists in national institutions to set research priorities, develop technologies suited to the specific needs of the region, and to evaluate and test the new technologies. There are currently 7,000 staff working at the international centers; 600 are senior scientists recruited from over 50 different countries. The CGIAR is formally cosponsored by the World Bank, the United Nations Development Program (UNDP), and the Food and Agriculture Organization (FAO). In 1984, the CGIAR will finance the 13 centers' collective budgets for a total of $180 million.

Contributions of the CGIAR System

Four of the institutes were initiated in the 1960's by the Ford and Rockefeller Foundations. The remainder were set up at different times since 1971 when the CGIAR was established. Despite their relatively short existence, the international centers have made remarkable contributions to world food production. As might be expected ,the impact of the two oldest institutes, the Centro Internacional de Mejoramiento de Maiz y Trigo (CIMMYT) in Mexico, and the International Rice Research Institute (IRRI) in the Philippines, has been the most dramatic. Their mandated crops, rice and wheat, are the two most widely consumed food crops in the world. High yielding, management responsive varieties of rice and wheat derived from those developed by CIMMYT and IRRI are grown on over 55 million hectares — one third of the area planted to these cereals in the developing world. It is estimated that the increased yields feed 300 million people and the economic value of this additional food supply is more than $5 billion annually. While less well known, the accomplishments of the newer centers are also making an impact on food production and hold even greater potential for the future. For example, in Latin America national programs have adopted and released more than 100 bean varieties developed by scientists at the Centro Internacional de Agricultura Tropical (CIAT) in Colombia. CIAT is also working on genetic improvements that can double and triple yields of cassava, an important food for some 400 million people in the tropics. In Lima, Peru, the Centro Internacional de la Papa (CIP) is concentrating on increasing production of the world's fourth most important crop. CIP houses the largest collection of potato germplasm in the world. A unique tissue culture technique makes possible the transmission of germ-free potato genetic resources. Improved lines of edible legumes and innovative soil and crop management systems for Asia and Africa are being developed by the International Crops Research Institute for the Semi-Arid Tropics (ICRISAT) based in India.

In addition to those international agricultural research centers supported by the CGIAR there are several funded independently of the Consultative Group. To further their objectives all the international centers are seeking more extensive exchange of knowledge on agricultural research and production among scientists and educators, particularly in developing countries. At the 1983 Frankfurt Bookfair they look forward to establishing new contacts and opening new channels of communication with publishers and book distributors from throughout the world.

Annual Reports

Consultative Group on International Agricultural Research Annual Report for 1984

October 1985. 68 pages. Perfect bound. Free (Annual Report for 1985, in press).

The CGIAR annual report provides the most comprehensive review of the CGIAR's current activities available anywhere to the general public. The 1984 and 1985 editions cover topics such as the impact of international agricultural research centers, research accomplishments, key events and CGIAR finances.

Monographs

Warren C. Baum

Partners against Hunger: The Consultative Group on International Agricultural Research

November 1986. 350 pages. 16 × 24 cm. Hard cover (ISBN 0-8213-0827-0), $29.95; paperback (ISBN 0-823-0828-9), $10.95. Plus air mail postage ($4.50 per item if mailed from Washington, D.C.); for surface mail, no charge (but all overseas orders are forwarded to local distributors whenever possible). Outside the U.S., this book may be purchased in local currency directly from local distributors of World Bank Books.

Describes how the CGIAR came into being, how it operates, and how it has forged a partnership of scientists and aid administrators from industrial and developing countries. Assesses the impact of the CGIAR on agricultural development, analyzes its successes and failures, and considers the potential for replication in other fields.

Donald L. Plucknett, Nigel J.H. Smith, J.T. Williams, and N.M. Anishetty

Gene Banks and the World's Food

January 1987. 248 pages. Hard cover. ISBN 08438-6. US$35.00 (gross price).

The authors address a broad audience of interested citizens, policy makers, and researchers to show why gene banks are emerging as linchpins in the global effort to conserve as much of the gene pool of crop plants as possible. The reader will find here a history of germplasm preservation and exchange, from botanical gardens to modern cold-storage units, and a highly readable assessment of the scientific, socioeconomic, and political ramifications of genebank programs. Gene Banks and the World's Food contributes to the crucial debate on how best to preserve some of society's most valuable raw material. The authors also provide an up-to-date report on the status and locations of gene banks, which includes the latest available information on germ-plasm holdings by crop. They then discuss how these holdings are being used to develop better crop varieties for the benefit of people around the world.

Miscellaneous

The Fragile Web: The International Agricultural Research System

1983. 27 pages. 17 × 25 cm. Saddle stitched. ISBN 0-88936-382-X. Free.

To highlight the first decade of research in the CGIAR system, Canada's IDRC — the International Development Research Centre — put together this publication outlining the work of the 13 CGIAR-affiliated centers. The booklet describes the origins of international agricultural research, the centers' current research agendas and future prospects for improving agricultural productivity in developing countries.

Robert S. McNamara

The Challenges for Sub-Saharan Africa

November 1, 1985. 49 pages. 14 × 22 cm. Saddle stitched. Free.

Mr. McNamara, a former President of the World Bank, states that the economic crisis in Sub-Saharan Africa threatens to "condemn an entire continent to unimaginable human misery" unless stronger action is taken to control population growth; reverse ecological devastation; eliminate distortions in domestic economic policies; and increase external development finance. As the first speaker in the Sir John Crawford Memorial Lecture Series, Mr. McNamara goes on to say that the current crisis in Africa requires a substantial increase in bilateral and multilateral aid, further debt rescheduling and the establishment of a special research program to increase the world's understanding of African ecological issues.

Robert S. McNamara

L'Afrique Subsaharienne: Un Devenir Difficile

November 1, 1985. 49 pages. 14 × 22 cm. Saddle stitched. Free.

Mr. McNamara, ancien President de la Banque Mondiale, écrit que la crise économique en Afrique sub-Saharienne menace de (condamner un continent entier à une misére humaine inimaginable), si on ne prend pas de mesures plus énergiques pour contrôler l'accroissement de la population, revenir sur la devastation ecologique, éliminer les distortions dans les politiques économiques, et augmenter le financement externe du développement. En tant que premier conférencier dans la série des Conferences Commémoratives de Sir John Crawford, Mr. McNamara poursuit en disant que la crise actuelle en Afrique necessite une augmentation substantielle de l'aide bilatérale et multilaterale, un réechelonnement supplémentaire de la dette, et la mise en oeuvre d'un programme de recherche spécial pour accrôitre la compréhension des problèmes de l'écologie africaine dans le monde.

Training in the CGIAR System

1986. 136 pages. 18.41 × 27.76 cm. Paperback. ISBN 971-104-161-8. Free.

The International Agricultural Research Centers emphasize training as an essential thrust to generate, promote, and disseminate research results. This publication describes efforts to forge and strengthen the partnership of the Consultative Group on International Agricultural Research and the nations it serves in its training endeavors to translate knowledge into more and better food from farmers' fields. It is based mainly on the report of a study team commissioned by the Technical Advisory Committee (TAC) to the CGIAR to review progress in training at the Centers and to suggest ways to further enhance this progress.

IRRI arranged the printing of the publication and assists the CGIAR and TAC in its distribution.

Impact Study Publications

In 1983, the CGIAR commissioned a wide-ranging study to assess the impact of the international agricultural research centers that fall under the CGIAR umbrella. As part of that exercise, a series of publications and papers were produced; the following list indicates those that are currently available and those scheduled for publication in 1986-87.

Main Report

(Untitled and currently being edited; estimated publication date mid-1987).

Summary of International Agricultural Research Centers: A Study of Achievements and Potential

30 pages. 28 × 21.5 cm. Free. Available from the CGIAR Secretariat.

Associated Reports

Dana G. Dalrymple, U.S. Agency for International Development

Development and Spread of High-Yielding Rice Varieties in Developing Countries

1986. US$12.75. Available from the International Rice Research Institute, P.O, Box 933, Manila, Philippines or Agribookstore, 1611 North Kent St., Arlington, Va. 22209, USA.

Dana G. Dalrymple, U.S. Agency for International Development

Development and Spread of High-Yielding Wheat Varietes in Developing Countries

1986. US$10.95. Available from CIMMYT, P.O. Box 6-641, Mexico 06600, D.F. Mexico or Agribookstore, 1611 North Kent St., Arlington, Va. 22209, USA.

Study Papers

(Available from the World Bank Book Store, 1818 H St., N.W. Washington, D.C. 20433; all study papers 21 × 27 cm).

Alain de Janvry and Jean-Jacques Dethier

Technological Innovation in Agriculture: The Political Economy of Its Rate and Bias

September 1985. ISBN 0-8213-0592-1. US$5.00.

Michael Lipton with Richard Longhurst

Modern Varieties, International Agricultural Research and the Poor

October 1985. ISBN 0-8213-0641-3. US$8.00.

J.G. Hawkes

Plant Genetic Resources: The Impact of the International Agricultural Research Centers

October 1985. ISBN 0-8213-0639-1. US$5.00.

Rigoberto Stewart

Costa Rica and the CGIAR Centers: A Study of Their Collaboration in Agricultural Research

September 1985. ISBN 0-8213-0615-4. US$5.00.

Rigoberto Stewart

Guatemala and the CGIAR Centers: A Study of Their Collaboration in Agricultural Research

October 1985. ISBN 0-8213 0616 2. US$5.00.

K.J. Billing

Zimbabwe and the CGIAR Centers: A Study of Their Collaboration in Agricultural Research

October 1985. ISBN 0-8213-0643-X. US$8.00.

Ramesh P. Sharma and Jock R. Anderson

Nepal and the CGIAR Centers: A Study of Their Collaboration in Agricultural Research

October 1985. ISBN 0-8213-0644-8. US$5.00.

Carl E. Pray and Jock R. Anderson

Bangladesh and the CGIAR Centers: A Study of Their Collaboration in Agricultural Research

October 1985. ISBN 0-8213-0642-1. US$5.00.

Fernando Homen de Melo

Brazil and the CGIAR Centers: A Study of Their Collaboration in Agricultural Research

January 1986. ISBN 0-8213-0694-4. US$5.00.

Barry Nestel

Indonesia and the CGIAR Centers: A Study of Their Collaboration in Agricultural Research

November 1985. ISBN 0-8213-0657-X. US$8.00.

Rafael Posada Torres

Ecuador and the CGIAR Centers: A Study of Their Collaboration in Agricultural Research

In press.

Luis J. Paz Silva

Peru and the CGIAR Centers: A Study of Their Collaboration in Agricultural Research

In press.

Hisham El-Akhrass

Syria and the CGIAR Centers: A Study of Their Collaboration in Agricultural Research

In press.

Pedro A. Sanchez & Grant M. Scobie

Cuba and the CGIAR Centers: A Study of Their Collaboration in Agricultural Research

In press.

Arturo A. Gomez

Philippines and the CGIAR Centers: A Study of Their Collaboration in Agricultural Research

In press.

Rungruang Isarangkura

Thailand and the CGIAR Centers: A Study of Their Collaboration in International Agricultural Research

In press.

Janice Jiggins

Gender Issues in International Agricultural Research: The Impact of International Agricultural Centers

In press.

Dev Bhumbla, Ishwar Mahapatra with Steve Bokil

India and the International Crops Research Institute for the Semi-Arid Tropics: A Study of Their Collaboration in Agricultural Research

In press.

Hans Jahnke

The Impact of Agricultural Research in Tropical Africa: A Study of the Collaboration Between the International and National Agricultural Research Systems

In press.

Eduardo Venezian

Chile and the CGIAR Centers: A Study of Their Collaboration in Agricultural Research

In press.

Kyaw Zin

Burma and the CGIAR Centers: A Study of Their Collaboration in Agricultural Research

In press.

Robert E. Evenson

The IARCs: Evidence of Impact on National Research and Extension Programs and on National Crop Yields

In press.

CIAT

Centro Internacional de Agricultura Tropical

Apartado Aéreo 6713 - Cali, Colombia
Cable: CINATROP - Telex 05769 CIAT Co
Telephone: 57-3-680111

General Information

Major research programs:
Cassava, field beans, rice, tropical pastures

Objectives:
To work with national organizations to develop improved agricultural technology to increase the quality and quantity of specific basic food commodities in the tropics. The main crops are cassava, field beans, and rice. CIAT also conducts a large program on developing suitable pasture technology for the infertile, acid soils of Latin America.

In this catalog CIAT publications are listed in the following criteria: General; Beans (*Phaseolus vulgaris*); Cassava; Pastures (tropical pasture species); Rice; Seeds (seed technology).

Order information:
As a nonprofit organization, CIAT is primarily interested in making its research information widely available to the people who need it. Prices for developing countries are determined from direct production costs plus a small markup to cover warehouse and distribution costs. The "list price" for each book is its commercial sale price, which is based on actual production costs. This is also the developed country cost for the book. (Discounts are available for educational institutions and for distribution agents.)

Libraries can request copies of the CIAT Report, Miscellaneous Reports, CIAT International, commodity newsletters, and the abstracts journals through the Library exchange System at CIAT.

Many publications are available in English through the IADS Agri-bookstore, Rosslyn Plaza, 1611 Kent Street, Suite 600, Arlington, West VA 22209, USA. (Price list available on request.)

Payment:
Prices are listed in US dollars. Payment may be made with a personal or cashiers check, drawn *only* on banks that are affiliated with a US bank. PLEASE NOTE THAT PAYMENT MUST ACCOMPANY THE ORDER.

Payment may also be made in Colombian pesos, or in CIAT coupons, AGRINTER coupons, or UNESCO coupons.

Checks should be made payable to CIAT. No refunds can be made for checks exceeding the exact amount of the invoice, due to currency restrictions. An invoice can be requested, in order to verify availability of materials and total price. All refunds will be made in CIAT coupons.

Send orders to:
>   Distribution Office
>   CIAT
>   Apartado Aéreo 6713
>   Cali, Colombia

Como hacer pedidos:
Al ser una organizacion sin animo de lucro, el CIAT esta principal-mente interesado en que la informacion sobre aus programas sea ampliamente conocida por aquellos que la necesitan. Los precios para paises en desarrollo reflejan los costos directos de publicacion, con un incremento pequeno para cubrir los costos de almacenamiento y distribucion.

Las bibliotecas pueden solicitar copias del Informe CIAT, de CIAT Internacional, de los boletines e, informes miscelaneos, y de las revistas de resumenes, mediante el sistema de canje del Centro.

Forma de pago:
Los precios anotados indican el precio en dolares estadounidenses por cada ejemplar, incluyendo el costo de su porte via correo aereo. SE REQUIERE PAGO ADELANTADO PARA CUALQUIER ENVIO DE PUBLICACIONES.

Se puede pagar con cheque personal o de gerencia en dolares estadounidenses y girado solamente contra bancos cuya oficina principal este localizada en los Estados Unidos. Si el pago se hace en pesos colombianos, por favor envie cheque giro o giro bancario.

El pago puede hacerse tambien con cupones del CIAT, cupones AGRINTER, y cupones de la UNESCO.

Los cheques deben ser girados a nombre del CIAT. Favor tomar nota de que no es posible devolver dineros sobre cheques recibidos cuyo valor exceda el total de la factura.
Envie los pedidos a la siquiente direccion:
>   Distribucion de Publicaciones
>   CIAT
>   Apartado Aéreo 6713
>   Cali, Colombia

Research Highlights

J. Reeves, Senior Writer

CIAT Report 1986/Informe CIAT 1986

1986. 140 pages. 21 × 27 cm. Perfect bound, paperback, ISSN 0120-3169 ISSN 0120-3150.

Highlights of the objectives, and research results in 1985 of CIAT's four basic crop commodity programs and support units.

Conferences

Federico Poey (Ed.)

Memorias Reunion de Trabajo Sobre Desarrollo y Proyeccion del Sector de Semillas en America Latina y El Caribe

1985. 128 pages. 15 × 21 cm. Paperback. ISBN 84-89206-47-3. HDC $6.00, LDC $5.00 plus airmail or surface mail postage.

Resumenes de los trabajos presentados sobre la situación del potencial de germoplasma disponible en los Centros sector de semillas en América Latina y el Caribe, el Internacionales de la región y de las posibilidades de cooperación técnica y financiera de las instituciones de apoyo que trabajan en el area.

Available only in Spanish.

Polania, Fabio (Ed.)

Memorias de la Reunion de Trabajo Sobre Investigacion y Capacitacion en Produccion y Tecnologia de Semillas

1986. 120 pages. 15 × 21 cm. Paperback. HDC $6.00, LDC $5.00 plus airmail or surface mail postage.

Resumenes de los trabajos presentados en la Reunión de Trabajo sobre Investigación y Capacitación en Tecnología de Semillas. Análisis del estado de la investigación sobre semillas, los logros alcanzados en este campo y las perspectivas que ofrecen nuevos campos de investigación sobre semillas en países desarrollados.

Disponible solo en Español.

John K. Lynam, Technical Editor

Cassava in Asia, its Potential and Research Development Needs

1986. 442 pages. 15 × 22 cm. Paperback. ISBN 84-89206-48-1. HDC $18.00, LDC $11.00 (airmail postage included).

Proceedings of a regional workshop held in Bangkok, Thailand, June 1984. Individual reports review the market potential of cassava in tropical Asia, new technology as the basis for increased production, and research and development needs.

J. Cock, Technical Editor

Global Workshop on Root and Tuber Crops Propagation

1986. 236 pages. 15 × 22 cm. Perfect bound, paperback. ISBN 84-89206-53-8. HDC $15.00, LDC $9.00 (airmail postage included).

Proceedings of a global workshop held in CIAT, Cali, Colombia in September 1983 covering crop description of potato, cassava, sweet potato, and cocoyam, seed production and storage and rapid propagation techniques. Sponsorship by CIAT, CIP, IITA, UNDP.

F. Cuevas and A.L. de Román (Editors)

Taller sobre la Red Cooperativa de Investigación de Arroz en el Caribe

1986. 138 pages. 15 × 22 cm. Perfect bound, paperback. ISBN 84-89206-57-0. HDC $15.00, LDC $9.00 (airmail postage included).

Proceedings of a workshop held in the Dominican Republic, August 1984. Eight country reports on the rice situation, production systems, research, extension and recommendations for future research; workshop recommendations including the structuring of a Caribbean rice research network.

C. Lascano and E. Pizarro (Editors)

Evaluación de pasturas con animales. Alternativas metodológicas

1986. 290 pages. 15 × 22 cm. Hardcover, perfect bound. ISBN 84-8280-154-6. HDC $15.00, LDC $10.00 (airmail postage included).

Proceedings of the workshop of the Red Internacional de Evaluación de Pastos Tropicales (RIEPT; International Network for Evaluation of Tropical Pastures) held in Lima and Pucallpa, Peru, October, 1984.

Miscellaneous (Thesaurus, Glossaries and Directories)

Anzola, C; Leatherdale,D; Youngberg, H.

Thesaurus on Seed Science and Technology

1986. 236 pages. 22 × 28 cm. Paperback. HDC $4.00, LDC $2.50 plus airmail or surface mail postage.

A list of terms related to Seed Science and Technology, presenting their synonyms, structure and relationships.

José Fernández de Soto

Glosario de Terminos Usados en Semillas

1986. 92 pages. 22 × 28 cm. Paperback. HDC $4.00, LDC $2.50 plus airmail or surface mail postage.

A list of terms most commonly used in Seed Science and Technology. Includes definition and English equivalences.

Available only in Spanish.

Anzola, C; Rodríguez, M.

Directorio de Specialistas e Instituciones en Semillas

1986. 121 pages. 22 × 28 cm. Paperback. Free.

A list of names, addresses and area of specialization for individuals and institutions involved with seed production and technology in Latin America and the Caribbean.

Technical Manual & Catalogs

R. S. Bradley et al.

Catálogo de cepas de rizobios para leguminosas forrajeras tropicales. 4a. ed.

Catalog of rhizobium strains for tropical forage legumes. 4th edition.

1986. 86 pages. 28 × 21 cm. Perfect bound. Paperback. Free.

Contains information related to the rhizobium strains collection kept at the Soil Microbiology section of CIAT's Tropical Pastures Program.

C.F. de Fernández, P. Gepts and M. López

Etapas de desarrollo de la planta de frijol común (*Phaseolus vulgaris* L.)

1986. 34 pages. 12 × 19 cm. Saddle stitched. ISBN 84-89206-54-6. HDC $9.00, LDC $7.00 (airmail postage included).

Illustrated description of the development stages and the vegetative cycle of the common bean plant.

Bibliographies

Centro Internacional de Agricultura Tropical
Bibliography on Bean Research in Africa
Supplement 1986

1986. 190 pages. 15 × 22 cm. Perfect bound, paperback. ISBN 84-89206-52-X. HDC $10.00, LDC $6.00 (airmail postage included).

This supplement includes 55 references to African bean literature. These documents, most difficult to obtain, give the history of research on *Phaseolus vulgaris* on the African Continent. Abstracts have been provided for 524 citations.

Cassava Information Center Staff
Abstracts on Cassava (*Manihot esculenta* Crantz) (Vol. XII)

1986. no. 1, 86 p; no. 2, 111 p; no. 3, 120 pages. 15 × 22 cm. Perfect bound, paperback. ISSN 0120-288X. HDC $35.00, LDC $16.00 plus air or surface mail postage.

Subject specialists analyze hundreds of articles in scientific and technical journals, manuals, bulletins, mimeographod works, and other sources and abstract the contents, augmenting the citation with keywords for quick retrieval. These abstracts are arranged by subject heading in periodic volumes that correspond to three key areas of CIAT research. All abstracts are subject and author indexed. Every title listed in the abstracts is available through the Documentation Service Unit, at CIAT.

Cassava Information Center Staff
Resumenes Analiticos Sobre Yuca (*Manihot esculenta* Crantz) (Vol. XII)

1986. no. 1, 119 p; no. 2, 118 p; no. 3, 120 pages. 15 × 22 cm. Perfect bound, paperback. ISSN 0120-2898. HDC $35.00, LDC $16.00 plus air or surface mail postage.

Especialistas en información analizan cientos de artículos publicados en revistas científicas y técnicas, manuales, boletines, trabajos mimeografiados y de otras fuentes, y resumen sus contenidos ampliando la cita bibliográfica con palabras claves que permiten una recuperación rápida. Estos resúmenes se organizan por encabezamiento de materia, en volúmenes periódicos que corrersponden a las 3 áreas claves de la investigación del CIAT. Todos los resúmenes se indizan por autor y por materia y cada título indicado en los resumenes esta disponible por intermedio de la Unidad de Servicios de Documentación del CIAT.

Available in English and Spanish.

Bean Information Center Staff

Abstracts on Field Beans (*Phaseolus vulgaris* L.) (Vol. XI)

1986. no.1, 190 p; no.2, 176 p; no.3, 190 pages. 15 × 22 cm. Perfect bound, paperback. ISSN 0120-2928. HDC $35.00, LDC $16.00 plus air or surface mail postage.

Subject specialists analyze hundreds of articles in scientific and technical journals, manuals, bulletins, mimeographed works, and other sources and abstract the contents, augmenting the citation with keywords for quick retrieval. These abstracts are arranged by subject heading in periodic volumes that correspond to three key areas of CIAT research. All abstracts are subject and author indexed. Every title listed in the abstracts is available through the Documentation Service Unit, at CIAT.

Bean Information Center Staff

Resumenes Analiticos Sobre Frijol (*Phaseolus vulgaris* L.) (Vol. XI)

1986. no.1, 198 p; no.2, 189 p; no.3, 196 pages. 15 × 22 cm. Perfect bound, paperback. ISSN 0120-2871. HDC $35.00, LDC $16.00 plus air or surface mail postage.

Documentalistas especializados analizan cientos de articulos publicados en revistas cientificas y técnicas, manuales, boletines, trabajos mimeografiados y de otras fuentes, y resumen sus contenidos ampliando la cita bibliográfica con palabras claves que permiten una recuperación rápida. Estos resúmenes se organizan por encabezamientos de materia, en volúmenes periódicos que corresponden a las 3 áreas claves de la investigación del CIAT. Todos los resúmenes se indizan por autor y por materia, y cada titulo indicado en los resúmenes esta disponible por intermedio de la Unidad de Documentación del CIAT.

Available in English and Spanish.

Pastures Information Center Staff

Resumenes Analiticos Sobre Pastos Tropicales (Vol. VIII)

1986. no. 1, 185 p; no.2, 121 p; no.3, 150 pages. 15 × 22 cm. Perfect bound, paperback. ISSN 0120-2944. HDC $35.00, LDC $16.00 (plus air or surface mail postage).

Especialistas en información analizan cientos de articulos publicados en revistas cientificas y técnicas, manuales, boletines, trabajos mimeografiados y de otras fuentes, y resumen sus contenidos ampliando la cita bibliográfica con palabras claves que permiten una recuperación rápida. Estos resúmenes se organizan por encabezamiento de materia, en volúmenes periódicos que corresponden a las 3 áreas claves de la investigación del CIAT. Todos los resumenes se indizan por autor y por materia y cada titulo indicado en los resúmenes está disponible por intermedio de la Unidad de Servicios de Documentación del CIAT.

(Subject specialists analyze hundreds of articles in scientific and technical journals, manuals, bulletins, mimeographed works, and other sources and abstract the contents, augmenting the citation with keywords for quick retrieval. These abstracts are arranged by subject heading in periodic volumes that correspond to three key areas of CIAT research. All abstracts are subject and author indexed. Every title listed in the abstracts is available through the Documentation Services Unit, at CIAT.)

Available in Spanish only.

A. Ramírez P. (Editor)

Pasturas tropicales - boletín. Vol. 8, Nos. 1-3

1986. 32 pagos. 21 × 27 cm. Saddle stitched. ISSN 0120-6915. Free.

Periodical bulletin with scientific articles, research activities and other information of interest to pasture researchers in the lowland tropics of Latin America and other areas in the tropics. It serves as a medium of communication to members of the Network for Evaluation of Tropical Pastures (RIEPT).

J. Reeves (Writer) / F. Motta (Editor)

CIAT Internacional. Vol. 5, Nos. 1-3

1986. 12 pages. 21 × 27 cm. Saddle stitched. ISSN 0120-4092. Free.

Newsletter. Highlights research results and international cooperation activities of CIAT. Published three times a year.

J. Reeves (Writer) / N. Molzan (Editor)

CIAT International. Vol. 5, Nos. 1-3

1986. 12 pages. 21 × 27 cm. Saddle stitched. ISSN 0110-4084. Free.

Newsletter. Highlights research results and international cooperation activities of CIAT. Published three times a year.

J. Reeves (Writer) / F. Motta (Editor)

Yuca Boletín Informativo. Vol. 10, Nos. 1-3

1986. 12 pages. 21 × 27 cm. Saddle stitched. ISSN 0120-1824. Free.

Highlights the main advances in research on this crop through contributions of researchers and specialists worldwide. Published three times a year.

J. Reeves (Writer) / Elizabeth L. de Páez (Editor)

Cassava Newsletter. Vol. 10, Nos. 1-3

1986. 12 pages. 21 × 27 cm. Saddle stitched. ISSN 0120-1824. Free.

Highlights the main advances in research on this crop through contributions of researchers and specialists worldwide. Published three times a year.

J. Reeves (Writer) / A.L. de Román (Editor)

Hojas de Frijol para América Latina. Vol. 8, Nos. 1-3

1986. 4 & 6 pages. 21 × 27 cm. Saddle stitched. ISSN 0120-2480. Free.

Newsletter with research activities and results of interest to members of bean research networks in Latin America. Published three times a year.

J. Reeves (Writer) / A.L. de Román (Editor)

Arroz en las Américas. Vol. 7, Nos. 1 & 2

1986. 8 pages. 21 × 27 cm. Saddle stitched. ISSN 0120-2634. Free.

Newsletter with research activities and results and news of interest to members of rice research networks in Latin America and rice researchers in general. Published twice a year.

Audiovisual Materials

C.A. Valencia

Benefico de semillas de *A. gayanus*

1986. 75 slides, 20 minutes.

Androgopon gayanus es una gramínea forrajera estable y productiva en las regiones de sabana de América tropical, por lo que ha sido liberada como cultivar en varios países de la región.

Las semillas de *A. gayanus* presentan problemas para el proceso de beneficio, debido a su bajo peso, abundante pubescencia y a la presencia de aristas y espiguillas vanas. Esta unidad presenta una descripción completa del proceso de beneficio, el cual puede producir tres clases de semilla de acuerdo con el grado de refinamiento: semilla cruda, semilla limpia o semilla clasificada.

C.A. Valencia

Main pests of stored bean

1986. 80 slides, 30 minutes.

Stored bean is attacked by pests commonly called bruchids that sometimes can cause irreparable losses.

This unit presents detailed information on the biology of the bruchids, and applicable methods for the control of these populations at domestic, small-farmer and commercial level. It also includes a description of the methodology used for their study.

CIP

**International Potato Center
Centro Internacional de la Papa
Centre International de la Pomme de Terre**

Apartado Postal 5969, Lima, Peru
Cables: CIPAPA-LIMA
Telex: 25672 PE
Telephone: 354354-366920

Foreword

The International Potato Center (CIP) is a scientific institution established for the purpose of developing and disseminating knowledge for greater use of the potato as a basic food. The following publications are designed to support research and communication among agricultural programs around the world. Several CIP publications are out of print and no longer listed. Copies may be available for consultation at an agricultural library.

We recommend contacting your National Potato Program. It is authorized to reproduce all technical information of CIP, and may also respond to other needs you may have concerning potato research or training.

Prefacio

El Centro Internacional de la Papa (CIP) es una entidad cientifica establecida para desarrollar y diseminar conocimientos sobre la papa, con el proposito de lograr su mayor utilizacion como alimento basico. Las siguientes publicaciones han sido hechas para apoyar la investigacion y comunicacion entre los programas agricolas en el mundo. Algunas publicaciones del CIP estan agotadas y ya no son incluides en esta lista. Podrian encontrarse disponibles en una biblioteca agricola.

Le recomendamos que se comunique con su Programa Nacional de Papa. Este esta autorizado para reproducir toda la informacion tecnica del CIP y quizas tambien pueda responder a otras necesidades que usted tenga en relacion con la investigacion y la capacitacion en papa.

Avant-propos

Le Centre International de la Pomme de Terre (CIP) est une institution scientifique qui a ete etablie dans le but de developper et disseminer les connaissances sur la Pomme de terre pour stimular son emploi en qualite d'aliment de base. Les publications presentees ci-apres ont ete faites afin d'encourager la recherche et la communication entre les programmes agricoles dans le monde entier. Quelques-unes des publications du CIP sont epuisees et ne sont plus distribuees par le CIP. Cependant, il se peut qu'elles soient disponibles dans d'autres bibliotheques specialisees.

Aussi, nous vous recommandons de contacter votre Programme National de Pomme de terre. Celui-ci est autorise a reproduire toute information technique venant du CIP et pourrait peut-etre repondre a d'autres besoins que vous pourriez avoir concernant la recherche et la formation en matiere de Pomme de terre.

Instructions for ordering

1 Mail your order form to the address most convenient to you (see pages 26 to 32).
2 CIP cannot assure that all publications are available at any time. We suggest to request proforma invoice before ordering. *Do not request* publications that are not listed or "in revision".
3 Pay with US$ check on a US bank or bank draft in US$ payable to CIP, or with Unesco coupons.

Note: Prices in this list include surface mail. Air mail on additional payment; see proforma invoice.

Procedimiento de compra

1 Envíe su orden de compra a la direccíon que más le convenga (vea las paginas 26-32)
2 El CIP no puede asegurarle que siempre tenga todas las publicaciones disponibles. Sugerimos solicitar factura proforma antes de hacer su pedido. *No solicite* material que no esté en la lista o que esté "en revisión".
3 Pague con cheque en US$ de un banco de los EE.UU. o giro bancario en US$ a nombre del CIP, o con bonos de la Unesco.

Nota: Los precios en esta lista incluyen el envío marítimo o terrestre. Vía aérea requiere pago adicional; ver factura proforma.

Comment passer votre commande

1 Envoyez votre formulaire de commande à l'adresse qui vous est la plus appropriée (voir les pages 26-32).
2 Le CIP ne peut pas vous assurer que toutes les publications soient toujours disponibles. Nous vous suggéront de solliciter une facture pro forma avant de passer votre commande. *Ne demandez pas* de publications marquées "en révision" ou qui ne figurent pas sur la liste.
3 Payez avec un chèque en US$ d'une banque des Etats-Unis ou un virement bancaire en US$ payable au CIP, ou avec des bons de l'Unesco.

Note: Les prix indiqués sur cette liste comprennent les frais d'envoi par bateau ou par voie de terre. Pour la voie aerienne, il y a un supplement; voir la facture pro forma.

Mail your order form to the address most convenient to you.
Envíe su formulario de pedido a la dirección que más le convenga.
Envoyez votre formulaire de commande a l'adresse la plus appropriée.

REGION I Latinoamérica Andina	CIP Apartado Aéreo 92654 Bogotá 8, D.E., *Colombia*
REGION II Latinoamérica no Andina	CIP c/o CNPH - EMBRAPA Caixa Postal (07)0218 70 359 Brasilia D.F., Brasil
REGION III Tropical Africa	CIP P.O. Box 25171 Nairobi, *Kenya*
REGION IV Middle East & North Africa	CIP P.O. Box 2416 Cairo, *Egypt*
REGION V Afrique du Nord et Afrique de l'Ouest	CIP 11 Rue des Orangers 2080 Ariana - Tunis, Tunisia
REGION VI South Asia	CIP c/o NBPGR IARI Campus New Delhi, 110012, *India*
REGION VII Southeast Asia	CIP c/o IRRI P.O. Box 933 Manila, *Philippines*
REGION VIII China	CIP c/o Chinese Academy of Agricultural Sciences Bai Shi Quio Rd. No. 30 Beijing, *P.R. of China*
CIP Main Headquarters Oficina Principal del CIP Direction Générale du CIP	CIP Apartado 5969 Lima, *Peru*

You can order certain CIP publications also through:
Usted también puede pedir ciertas publicaciones del CIP a:
Vous pouvez commander certaines publications du CIP aussi auprès
de:

Agribookstore Winrock International 1611 North Kent St. Arlington, VA 22209 USA	Phone 703-525-9455
Josef Margraf Eichendorfstr. 9 8074 Gaimersheim West Germany	Tel. 08458/2207
Editorial Agropecuaria Hemisferio Sur S.R.L. Casilla de Correo 1755 Montevideo Uruguay	Tel. 954454

New Publications	Nuevas Publicaciones	Nouvelles Publications
Informe Anual CIP 1984 Centro Internacional de la Papa. 1985. 186 pp.		free
(Annual Report CIP 1984		Out of print)
Innovative methods for propagating potatoes International Potato Center. 1985. 350 pp. Planning conference Report 28		US$ 5.-
Field-screening CIP clones developed for resistance to bacterial wilt French, E.R. 1982. 9 pp. Technology Evaluation Series No. 6		US$ 1.-
Evaluación de tecnología para la producción de tubérculos-semillas de semilla botánica de papa Wiersema, S.G. 1983. 14 pp. Serie de Evaluación de Tecnología No. 8		US$ 1.-
Evaluación de tecnología agronómica para producción de papa a partir de semilla botánica Malagamba, J.P. 1983. 20 pp. Serie de Evaluación de Tecnología No. 9		US$ 1.-

Evaluation of technology for integrated control
of potato tuber moth in field and storage
Raman, K.V., Booth, R.H. 1983. 18 pp.
Technology Evaluation Series No. 10

US$1.-

Evaluación de tecnología
para control integrado de la polilla de la papa
en campos y almacenes
Raman, K.V., Booth R.H. 1986. 30 pp.
Serie de Evaluación de Tecnología No. 12

US$ 1.-

Investigaciones nematológicas
en programas latinoamericanos de papa
Franco, J., Rincón, H. (Editores). 1985.
2 Volumenes, 189 pp.

US$ 6.-

Breaking new ground: Agricultural anthropology
Rhoades, R.E. 1984. 84 pp.

US$ 5.-

Tissue culture micropropagation, conservation
and export of potato germplasm
Espinoza, N., Estrada, R., Tovar, P., Bryan, J.,
Dodds, J.H. Revised 1986. 20 pp.

US$ 1.-

Cultivo de tejidos:
Micropropagación, conservación y exportación
de germoplasma de papa
Espinoza, N., Estrada, R., Tovar, P., Bryan, J.,
Dodds, J.H. 1985. 17 pp.

US$ 1.-

Culture de Tissus:
Micropropagation, conservation et exportation
du germoplasme de la pomme de terre
Espinoza, N., Estrada, R., Tovar, P., Bryan, J.,
Dodds, J.H. 1986. 22 pp.

US$ 1.-

Potato storage for developing countries
Booth, R.H., Shaw, R.L., Harmsworth, L.J.
1981. 24 pp.

US$ 1.-

Introduction to potato storage
Shaw, R.L., Booth, R.H. 1981. 10 pp.

US$ 1.-

Introducción al almacenamiento de la papa
Shaw, R.L., Booth, R.H. 1986. 10 pp.

US$ 1.-

Traditional potato production and farmers'
selection of varieties in eastern Nepal
Rhoades, E.R. 1985. 52 pp.
Potatoes in Food Systems Research Series No. 2

US$ 1.-

Household food production: US$ 1.-
Comparative perspectives
Niñez, V.K. (Editor). 1985. 67 pp.
Reprinted from Food and Nutrition Bulletin,
1985, Vol. 7, No. 3.

Catálogo de publicaciones periódicas   US$ 1.-
de la biblioteca del CIP
Centro Internacional de la Papa. 1986. 41 pp.

Markets, myths and middlemen: US$ 9.-
A study of potato marketing in central Peru
Scott, G.J. 1985. 196 pp.

Mercados, mitos e intermediarios: Contacte
La comercialización de la papa en la zona   el CIUP
central del Perú
Scott, G.J. 1985. 308 pp.
Centro de Investigación (CIUP)
Universidad del Pacífico
Avenida Salaverry 2020
Lima 11, Perú

Evaluation manual for CIP courses;   US$ 2.-
objectives and implementation procedures
Siri, C. 1986. 60 pp.

CIP Circular - Circular del CIP

Distribution on mailing list   free
Distribucion por subscripción   gratuita

Main topics of the past issues:
Temas principales de los últimos números:

Vol. 13, No. 1, 1985
• Production of seed potatoes
   derived from true seed
• Producción de tubérculos-semillas
   derivados de la semilla

Vol. 13, No. 2, 1985
• Basic seed: A new production
   and distribution program for Peru
• Semilla básica: Un nuevo programa de producción
   y distribución para el Perú

Vol. 13, No. 3, 1985
* The quest for improved potato varieties   Out of print
  in Thailand
* La búsqueda de variedades mejoradas de papa
  en Tailandia

Vol. 13, No. 4, 1985
* Induction and use of in vitro potato tubers
* Inducción y utilización de tubérculos de papa
  producidas *in vitro*

Working Papers - Documentos de Trabajo   US$ 1.-

1984 - 3 Building an effective potato country pro-
         gram:
         The case of Rwanda
         Monares, A. 34 pp.

1985 - 1 Adopción y difusión de variedades de papa
         en el departamento de Cajamarca
         Franco, E., Schmidt, E. 30 pp.

1985 - 2 Eficiencia técnica y económica
         de la producción de semilla de papa
         en la sierra central del Perú
         Auroi, C., Vilca, P. 49 pp.

Technical Information Bulletins (TIBs)   US$ 1.-
Boletines de Información Técnica (TIBs)
Bulletins d'Information Technique (TIBs)

English - The TIBs contain information that is useful for potato
         production and applied research. Although the information is
         directed at an intermediate professional level, it is adaptable to
         the farmers' level. The TIBs may be used for study, potato
         production, experimentation, training, and production of
         information for farmers.

Español - Los TIBs contienen información que es útil para la produc-
         ción de papa y la investigación aplicada. Aunque la informa-
         ción está dirigida a un nivel profesional intermedio, es
         fácilmente adaptable al nivel del agricultor. Los TIBs pueden
         ser usados para el estudio, la producción de papa, experi-
         mentación, capacitación, y elaboración de mensajes para los
         agricultores.

Nota: El CIP autorizó a la Editorial Agropecuaria Hemisferio Sur para
         reproducir los TIBs y distribuirlos en los países de lengua
         española de Suramérica, México y España. Sus precios varían
         (entre US$ 0.30 y 0.40 por TIB). Recomendamos escribir a la
         dirección especificada en la página 5 de este suplemento.

Francais - Les TIBs 1 a 19 ont été traduits en français et le CIP cherche actuellement un co-éditeur intéresse par leur édition et distribution sur le marche.

Hindi - Contact: CIP Region VII; address on page 26.

Bengali - Contact: Lyle C. Sikka, BARC, c/o Winrock International, Farm Gate Complex, Dhaka - 15, Bangladesh.

Farsi - Contact: Dr. Dariush Danesh, University of Isfahan, P.O. Box 156, Isfahan 81745, Iran.

Urdu - Contact: PAK - Swiss Potato Development Project, P.O. Box 1031, Islamabad, Pakistan.

Chinese - Contact: CIP Region VIII; address on page 26.

INTERNATIONAL POTATO CENTER (CIP)

PRO FORMA INVOICE	FACTURA PRO FORMA	FACTURE PRO FORMA	
Date	Fecha	Date: _______	
Your reference	Su referencia	Votre référence: _______	

Quantity / Cantidad / Quantité	Description / Descripción / Description	Unit / Unidad / Unité	Total
Publications	Publicaciones	Publications	US $ _____ *
Air mail	Envío aéreo	Frais d'envoi	US $ _____
TOTAL			US $ _____

*Price of publications include surface mail. Air mail extra.
*Los precios de las publicaciones incluyen el envío marítimo o terrestre. El envío aéreo requiere pago adicional.
*Les prix des publications comprennent les frais d'envoi par bateau ou par voie de terre. Il y a un supplément pour l'envoi par avion.

IMPORTANT See "Instructions for Ordering" in CIP's Publication List. Pay with US$ check on a US bank, or bank draft in US$, payable to CIP, and enclose this proforma invoice. We will send materials and receipt as soon as we receive payment.

IMPORTANTE Vea "Procedimiento de Compra" en la Lista de Publicaciones del CIP. Pague con cheque en US$ de un banco de los EE.UU., o giro bancario en US$, a nombre del CIP, e incluya esta factura proforma. Despacharemos los materiales y un recibo oficial tan pronto recibamos el pago.

IMPORTANT Voir "Comment passer votre commande" dans la liste des publications du CIP. Payez avec un cheque en US$ d'une banque des Etats-Unis, ou un virement bancaire en US$, payable au CIP. Dès réception du paiement, nous delivrerons une facture et enverrons le matériel.

Sender   Remitente   Expéditeur

____

____

____

____

____

CIMMYT

Centro Internacional de Mejoramiento de Maiz y Trigo (International Maize and Wheat Improvement Center)

Lisboa 27 - Apdo. Postal 6-641 - 06600 Mexico D.F.
Cable Address: CENCIMMYT - Telex address: 1772023 CIMTME
Telephone: 761-33-11; 761-36-56

General Information

Maize and wheat constitute the principal sources of carbohydrates and protein for nearly half of the people of the world. CIMMYT's primary objective is to develop superior germplasm that will provide higher and more stable yields, as well as better nutritional quality. Toward this end, CIMMYT promotes and helps to implement research, training, and information programs designed to improve maize and wheat production in the developing countries of the world. The Center currently conducts research on maize, bread wheat, durum wheat, and triticale.

To be consistent with its mandate to serve the Third World, CIMMYT provides single copies of most of its publications *free of charge* to over 5,000 agricultural scientists and libraries in developing countries. Beyond this complementary distribution, CIMMYT employs a two-tiered pricing structure for its publications. Separate prices are listed in this catalog for Highly Developed Countries (HDCs) and Less Developed Countries (LDCs). All prices include air mail postage. (Note: CIMMYT reserves the right to adjust prices on an annual basis.)

In addition, all CIMMYT publications are available on microfiche. If interested in this delivery format, please write for cost information and other details.

Research Highlights

CIMMYT Staff

CIMMYT Research Highlights 1985 (new listing)

1986. 116 pages. 22 × 25 cm. Paperback. Perfect bound. Available in English and Spanish. HDC $8.00. LDC $3.00.

An annual research report on selected CIMMYT research activities in maize, bread wheat, durum wheat, and triticale crop improvement. Highlights of selected crop management and economics research activities are also provided.

Also available: *Research Highlights* 1984

Annual Reports

CIMMYT Staff

CIMMYT Annual Report 1985

1986. 74 pages. 21.5 × 28 cm. Paperback. Saddle stitched. Available in English and Spanish. HDC $8.00. LDC $3.00.

This non-technical annual report is intended primarily for policy makers and members of the donor community. It presents CIMMYT's research program activities in a highly concise form (the "1985 Management Report" and "A Review of CIMMYT Programs"), and gives emphasis to the financial and special project aspects of the Center's operations. The *Annual Report* is supplemented by the *Research Highlights* and two technical reports, one focusing on maize research and the other on wheat research.

CIMMYT Maize Program Staff

CIMMYT Biennial Report on Maize Improvement 1982-83

Available January 1987. 21.5 × 28 cm. Paperback. Perfect bound. Available in English. HDC $8.00. LDC $3.00.

This technical report is produced at two-year intervals and is intended for informed, technically oriented individuals working on various aspects of maize improvement. The report describes in detail the CIMMYT Maize Program's research in gene pool development, population improvement, special projects, wide crosses, training and regional programs.

CIMMYT Wheat Program Staff

CIMMYT Report on Wheat Improvement 1984

1986. 21.5 × 28 cm. Paperback. Perfect bound. 186 pages. Available in English. HDC $8.00. LDC $3.00.

This annual technical report is intended for informed, technically oriented individuals working on various aspects of wheat improvement. The report describes the CIMMYT Wheat Program's research on bread wheat, durum wheat, and triticale. Also included are reports on activities of the research support programs, regional programs, and bilateral programs.

Also available: *CIMMYT Report on Wheat Improvement* 1983, 1982

Conference Proceedings

B. Gelaw, Editor

To Feed Ourselves: A Proceedings of the First Eastern, Central and Southern Africa Regional Maize Workshop

1986. 17 × 24 cm. Paperback. Perfect bound. Available in English. HDC $12.00. LDC $5.00.

A regional workshop on maize in Eastern, Central, and Southern Africa was held in Lusaka, Zambia, on March 10-17, 1985, and was attended by maize breeders and agronomists from 17 African countries. The proceedings contains "country reports" that summarize the status of maize production in the 17 countries represented at the workshop, as well as technical papers by selected specialists. These technical papers present and discuss 1) various approaches to maize research, 2) the constraints to maize production in African environments, 3) the breeding efforts necessary to overcome these constraints, 4) seed production issues, 5) questions regarding various economic aspects of mazie production under the conditions of representative farmers.

CIMMYT Staff

Strengthening Agricultural Research in Latin America and the Caribbean (Proceedings)

1985. 177 pages. 17 × 24 cm. Paperback. Perfect bound. Available in English and Spanish. HDC $12.00. LDC $5.00.

The Workshop on Strengthening Agricultural Research in Latin America and the Caribbean, held at CIMMYT on September 10-12, 1984, brought participants together to address issues related to the framework within which agricultural research is organized in these regions. The papers in the proceedings include a projection of production, consumption, and export trade trends in Latin America to the year 2000; a report on agricultural research in the public sector; and a summary of current opportunities for investment in agricultural research. On-farm research, biotechnology, and the availability of human resources for national agricultural research also receive attention. Observations of commentators are included.

Miscellaneous Reports and Brochures

CIMMYT Staff

Improving on Excellence: Achievements in Breeding with the Maize Landrace Tuxpeño

1986. 10.2 × 21.5 cm. Paperback. Saddle wire stitched. Available in English and Spanish. HDC $2.00. LDC $1.00.

Over the past two decades at CIMMYT, genetic improvement of Tuxpeño, one of the outstanding maize germplasm complexes of Latin America, has resulted in the development of materials that are useful in many parts of the world. The Program's approach to developing these materials provides a focus for describing several important aspects of its work. Among the aspects considered are the assumptions and decisions made by maize breeders about their work, goals and priorities they set, methods they applied, and results they achieved.

The progress of the Maize Program and indications of its future course are also discussed.

CIMMYT Staff

Veery "S": A Bread Wheat for Many Environments

1986. 10.2 × 21.5 cm. Paperback. Saddle wire-stitched. Available in English and Spanish. HDC $2.00. LDC $1.00.

This publication describes the Wheat Program's development of improved bread wheats from crosses between the spring-and winter-habit germplasm pools. Veery "S" materials have moved bread wheat onto a new yield level, and show exceptionally stable performance across a wide range of environments. Varieties derived from Veery "S" are being rapidly released by national programs, and are currently planted on over 3,000,000 ha worldwide.

Christopher R. Dowswell

Strengthening National Research Programs Through Training: A Twenty-Year Progress Report

1986. 15.2 × 28 cm. Paperback. Saddle wire stitched. Available in English and Spanish. HDC $5.00. LDC $2.00.

The evolution, organization, and philosophy of CIMMYT training are reviewed in this publication, with emphasis on the Center's contribution to expanding the capacity of national programs to conduct research in maize and wheat improvement and economics. The full range of training opportunities offered by CIMMYT is described in detail. The program's plans for the future, especially with regard to the changing needs of national programs, are discussed.

Steven A. Breth

Mainstreams of CIMMYT Research: A Retrospective

1986. 44 pages. 21.5 × 28 cm. Paperback. Saddle wire stitched. Available in English and Spanish. HDC $7.00. LDC $3.00.

CIMMYT's most significant research initiatives and accomplishments are reviewed in this publication, which includes special sections on the Center's origins and organization, its objectives and priorities during the past two decades, its work with national crop improvement programs, and its three major research programs — Maize, Wheat, and Economics. A comprehensive summary of the work of each of these programs to date, as well as the present direction of their research, provides a general perspective on CIMMYT's evolution as an organization.

CIMMYT Staff

Poster: Identification of Rust Diseases on Wheat

1985. 28 × 60.5 cm. Full color. Available in English, Spanish and French. HDC $2.00. LDC $1.00.

The three rust dieases occurring on wheat — stem rust, leaf rust, and stripe rust — are pictured to aid identification. A description of symptoms and disease development, as well as spore morphology, is given for each rust disease.

CIMMYT Staff

Poster: Characteristics of Selected Seed-borne Fungi

1985. 28.5 × 80 cm. Full color. Available in English, Spanish, and French. HDC $2.00. LDC $1.00.

Four seed-borne fungal diseases are compared to healthy seed: Karnal bunt, common bunt, dwarf bunt, and black point. For each disease, a description of the crops affected, the symptoms, and the spore morphology is given.

Guidebooks and Manuals

J.M. Prescott et al.

Wheat Diseases and Pests: A Guide for Field Identification

1986 (Third Edition). 135 pages. 106 color photographs. 10 × 18 cm. Paperback. Saddle wire stitched. Available in English and Spanish. HDC $10.00. LDC $4.00.

An illustrated, pocket-size field guide to common wheat diseases and pests, designed for use in the field by all agricultural workers. A simple diagnostic key, numerous color photographs of diseased and pest-infested plants, and a helpful glossary of pathology-related terms are included as aids to identification.

Periodicals

CIMMYT Economics Program Staff

1986 World Maize Facts and Trends: An Analysis of the Economics of Maize Seed Production

Available November 1986. 21.5 × 28 cm. Paperback. Saddle wire stitched. Available in English. HDC $8.00. LDC $3.00.

A viable, effective maize seed industry is a key element in the extension of improved maize varieties to Third World farmers. In this report, the worldwide use of maize seed is analyzed by major seed type, and the prices and costs of different types of maize seed are detailed. A discussion of issues central to maize seed production focuses on the use of open-pollinated maize seed and hybrid maize seed, on public and private seed industries, and on incentives for establishing an effective seed industry. In tabular form, the report presents recent data related to the maize economy in countries that either grow over 100,000 hectares of maize or consume 100,000 tons of the grain (or both).

CIMMYT Staff

A Common Ground for Maize Research: Regional Cooperation in the Middle East and North Africa. CIMMYT Today

1986. 19 pages. 21.5 × 28 cm. Saddle wire stitched. Full color presentation. Available in English and Spanish. IIDC $3.00. LDC no charge.

CIMMYT's Middle East/North Africa Regional Maize Program is tailored to the unique requirements and circumstances of maize research in the region, reinforcing the efforts of national researchers to improve the productivity of resources committed to maize production. Some of the accomplishments of these researchers in Egypt, Turkey, Morocco, Syria, and Portugal are described, including their work in germplasm development and on-farm research. This publication also explains how they are supported by CIMMYT's maize program, which develops and improves germplasm and delivers this and other research products or services to national maize scientists.

IBPGR

International Board for Plant Genetic Resources

Crop Genetic Resources Centre Food and Agriculture Organization of
the United Nations, Via delle Terme de Caracalla, 00100 Rome, Italy
Cable:FOODGARI, ROME - Telex: 610181 FAOI

Annual Reports, Periodicals and Research Highlights

IBPGR Annual Report 1985

1986. 92 pages. 19 × 26 cm. Paper. ISBN 92-9043-121-0. Available free to developing countries but restricted distribution to developed countries.

The Annual Report provides a detailed overview of IBPGR's activities during the year.

FAO/IBPGR Plant Genetic Resources Newsletter (Nos. 1-65).

Quarterly. Approx. 40 pages. 21 × 29.7 cm. Paper. Available free to developing countries, but restricted distribution to developed countries.

This newsletter describes the diverse activities of the global programme developed by the IBPGR and organized co-jointly with FAO.

L.A. Withers and J.T. Williams

Research Highlights: In Vitro Conservation

1986. 21 pages. 14.5 × 20.7 cm. Paper. Available free to developing countries, but restricted distribution to developed countries.

Up to 1984, all IBPGR-supported research had been summarized in Annual Reports. For 1984-85, IBPGR research on *in vitro* conservation forms the topic for this, the first IBPGR Research Highlights. Included are concepts and practices of *in vitro* conservation and IBPGR research on specific crops.

General Information

Crop Germplasm Conservation and Uses (Pamphlet)

1986. 13 pages. 10 × 21 cm. Paper. Available free to developing countries, but restricted distribution to developed countries.

This pamphlet briefly describes the uses put to conserved germplasm, particularly by plant breeders and notes new technologies which may be of use in the improvement of crop plants.

Monographs on Crops

J.C. Clemént

Les Mils Penicillaires de l'Afrique de l'Ouest: Prospections et Collectes

1985. 231 pages. 21 × 29.7 cm. Paper. Published in cooperation with Institut Francais de Recherche Scientifique pour le Developpement en Cooperation (ORSTOM). Available free to developing countries, but restricted distribution to developed countries.

This publication includes information on the variability in pearl millet (*Pennisetum americanum* L.) in the Sahelian countries obtained in collecting missions since the early 1970s. In French.

S.K. Mukherjee

Systematic and Ecogeographic Studies on Crop Genepools: 1. *Mangifera* L.

1985. 86 pages. 21 × 29.7 cm. Paper. Available free to developing countries, but restricted distribution to developed countries.

This is the first of a series which summarize studies of the geographical distribution of species, in this case those related to the cultivated mango with the purpose of devising a plan of action for collection and conservation (*in situ* and *ex situ*).

C.D.G. Chapman

Genetic Resources of Wheat: A Survey and Strategy for Collecting

1985. 39 pages. 21 × 29.7 cm. Paper. Available free to developing countries, but restricted distribution to developed countries.

Wheat germplasm has been extensively collected for over a relatively long period. This germplasm is dispersed among a number of collections however and the objective of this survey is get a coherent picture in order to set priorities for future work.

Genetic Resources of Tropical and Sub-Tropical Fruits and Nuts (Excluding *Musa*).

1986. 160 pages. 21 × 29.7 cm. Paper. Available free to developing countries, but restricted distribution to developed countries.

This book holds reports on specific aspects relevant to the conservation of genetic resources of 31 mainly tropical and sub-tropical fruit and nut crops. Some major crops such as banana and plantain, which have been covered in other IBPGR publications, are not included.

Monographs on Conservation

The Potential for Using *In Vitro* Techniques for Germplasm Collection

1984. 8 pages. 21 × 29.7 cm. Paper. Available free to developing countries, but restricted distribution to developed countries.

This book is an outcome of a meeting of a subcommittee of the IBPGR's Advisory Committee on *In Vitro* Conservation. It considers novel collecting techniques for vegetatively propagated species and material such as coconut which present practical problems.

J. Hanson

Procedures for Handling Seeds in Genebanks (Practical Manuals for Genebanks No.1)

1985. 115 pages. 21 × 29.7 cm. Spiral-bound. Available free to developing countries, but restricted distribution to developed countries.

This manual is the first in a series of practical guides to suitable methods for processing and storing seeds in genebanks. It is aimed specifically for genebank personnel, especially those technicians and staff involved in day-to-day seed handling.

Long-term Seed Storage of Major Temperate Fruits

1985. 38 pages. 21 × 29.7 cm. Paper. Available free to developing countries, but restricted distribution to developed countries.

This report provides information on the storage and germination of seeds of *Prunus, Malus, Pyrus* and *Vitis* spp. The problem of substantial seed dormancy is common to all four genera.

Cost-effective Long-term Seed Storage

1985. 38 pages. 21 × 29.7 cm. Paper. Available free to developing countries, but restricted distribution to developed countries.

This is a report of a meeting of a subcommittee of the IBPGR Advisory Committee on Seed Storage. Particular emphasis is given to alternatives to conventional long-term seed stores operating at the IBPGR preferred conditions.

C. Stushnoff and C. Fear

The Potential Use of *In Vitro* Storage for Temperate Fruit Germplasm: A Status Report.

1985. 21 pages. 21 × 29.7 cm. Paper. Available free to developing countries, but restricted distribution to developed countries.

This report discusses conservation of temperate fruit germplasm *in vitro*, either using minimum growth methods for short- to medium-term storage, or cryopreservation for long-term storage.

The Design, Planning and Operation of *In Vitro* Genebanks

1986. 17 pages. 21 × 29.7 cm. Paper. Available free to developing countries, but restricted distribution to developed countries.

This is a report of a meeting of a subcommittee of the IBPGR Advisory Committee on *In Vitro* Storage. It draws attention to the scale, in terms of manpower requirements and other resources, of undertaking the establishment of an *In Vitro* Active Genebank.

R.H. Ellis, T.D. Hong and E.H. Roberts

Handbook of Seed Technology for Genebanks Vol. I: Principles and Methodology (Handbooks for Genebanks No. 2)

1985. 210 pages. 21 × 29.7 cm. Paper. Available free to developing countries, but restricted distribution to developed countries.

This first volume on seed technology for genebanks deals with many of the principles of seed testing which need to be understood when monitoring the viability of seed accessions maintained in genebanks. It is the second in a general series of genebank handbooks.

R.H. Ellis, T.D. Hong and E.H. Roberts

Handbook of Seed Technology for Genebanks Vol. II: Compendium of Specific Germination Information and Test Recommendations (Handbooks for Genebanks No. 3)

1985. 456 pages. 21 × 29.7 cm. Paper. Available free to developing countries, but restricted distribution to developed countries.

This second volume on seed technology for genebanks provides general approaches, detailed information, guidance and, where available, prescriptions for removing dormancy and germinating the seeds. It is the third in a general series of genebank handbooks.

Documentation of Genetic Resources (General)

J. Konopka and J. Hanson (editors)

Documentation of Genetic Resources: Information Handling Systems for Genebank Management

1985. 87 pages. 17 × 24.5 cm. Paper. Available free to developing countries, but restricted distribution to developed countries.

This publication, the proceedings of a workshop, has the objective of raising the standards of genebank management by stimulating the development and implementation of computerized data management systems in order to facilitate seed handling procedures.

M.J.A. Simpson and L.A. Withers

Documentation of Genetic Resources: Characterization of Plant Genetic Resources Using *Isozyme Electrophoresis:* A Guide to the Literature

1986. 102 pages. 17 × 24.5 cm. Paper. Available free to developing countries, but restricted distribution to developed countries.

Morphological and agronomic evaluation of variability of accessions held in genebanks may be supplemented by more direct study of the genome by *isozyme electrophoresis.* This survey aims to provide a guide to the literature on isozyme electrophoresis in plant biology and to demonstrate the potential role of the techniques in the evaluation and utilization of plant genetic resources.

Descriptor Lists

This description applies to all the books in the following section: In order for the global network of crop genetic resources centres to readily exchange data about samples along with the plant materials, it is necessary for each centre to develop a data bank with a certain degree of standardization. This standardization is provided by the descriptors for each crop agreed internationally. The use of these descriptors permits scientists of different nationalities to readily communicated among themselves.

Forage Legumes Descriptors

1984. 29 pages. 21 × 29.7 cm. Paper. Published in cooperation with Commission of European Communities (CEC). Available free to developing countries, but restricted distribution to developed countries.

E. Bellini, R. Watkins and E. Pomarici (editors)

Peach Descriptors

1985. 34 pages. 21 × 29.7 cm. Paper. Published in cooperation with Commission of European Communities (CEC). Available free to developing countries, but restricted distribution to developed countries.

R. Guerriero and R. Watkins (editors)

Apricot Descriptors

1985. 36 pages. 21 × 29.7 cm. Paper. Published in cooperation with Commission of European Communities (CEC). Available free to developing countries, but restricted distribution to developed countries.

D. Cobianchi and R. Watkins (editors)

Plum Descriptors

1985. 31 pages. 21 × 29.7 cm. Paper. Published in cooperation with Commission of European Communities (CEC). Available free to developing countries, but restricted distribution to developed countries.

Descriptors for Finger Millet

1985. 20 pages. 17 × 24.5 cm. Paper. Available free to developing countries, but restricted distribution to developed countries.

Faba Bean Descriptors

1985. 19 pages. 17 × 24.5 cm. Paper. Published in cooperation with International Center for Agricultural Research in the Dry Areas (ICARDA). Available free to developing countries, but restricted distribution to developed countries.

Lentil Descriptors

1985. 15 pages. 17 × 24.5 cm. Paper. Published in cooperation with International Center for Agricultural Research in the Dry Areas (ICARDA). Available free to developing countries, but restricted distribution to developed countries.

Descriptors for *Vigna aconitifolia* and *V. trilobata*

1985. 39 pages. 17 × 24.5 cm. Paper. Available free to developing countries, but restricted distribution to developed countries.

Descriptors for *Setaria italica* and *S. pumila*

1985. 18 pages. 17 × 24.5 cm. Paper. Available free to developing countries, but restricted distribution to developed countries.

Descriptors for *Panicum miliaceum* and *P. sumatrense*

1985. 14 pages. 17 × 24.5 cm. Paper. Available free to developing countries, but restricted distribution to developed countries.

B.F. Tyler, J.D. Hayes and W. Ellis Davies (editors)
Forage Grass Descriptors

1985. 30 pages. 21 × 29.7 cm. Paper. Published in cooperation with Commission of European Communities (CEC). Available free to developing countries, but restricted distribution to developed countries.

Descriptors for *Phaseolus acutifolius*

1985. 26 pages. 17 × 24.5 cm. Paper. Available free to developing countries, but restricted distribution to developed countries.

Sunflower Descriptors

1985. 33 pages. 17 × 24.5 cm. Paper. Available free to developing countries, but restricted distribution to developed countries.

Chickpea Descriptors

1985. 15 pages. 17 × 24.5 cm. Paper. Published in cooperation with International Center for Agricultural Research in the Dry Areas (ICARDA) International Crops Research Institute for the Semi-Arid Tropics (ICRISAT). Available free to developing countries, but restricted distribution to developed countries.

Oat Descriptors

1985. 21 pages. 17 × 24.5 cm. Paper. Available free to developing countries, but restricted distribution to developed countries.

Cherry Descriptors

1985. 33 pages. 21 × 29.7 cm. Paper. Available free to developing countries, but restricted distribution to developed countries.

Cotton Descriptors (Revised)

1985. 17 pages. 17 × 24.5 cm. Paper. Available free to developing countries, but restricted distribution to developed countries.

Descriptors for Wheat (Revised)

1985. 12 pages. 17 × 24.5 cm. Paper. Published in cooperation with Commission of European Communities (CEC). Available free to developing countries, but restricted distribution to developed countries.

Descriptors for *Vigna mungo* and *V. radiata* (Revised)

1985. 23 pages. 17 × 24.5 cm. Paper. Available free to developing countries, but restricted distribution to developed countries.

Groundnut Descriptors (Revised)

1985. 20 pages. 17 × 24.5 cm. Paper. Published in cooperation with International Crops Research Institute for the Semi-Arid Tropics (ICRISAT). Available free to developing countries, but restricted distribution to developed countries.

R. Gülcan (editor)
Almond Descriptors (Revised)

1985. 30 pages. 21 × 29.7 cm. Paper. Published in cooperation with Commission of European Communities (CEC). Available free to developing countries, but restricted distribution to developed countries.

Rye and Triticale Descriptors

1985. 13 pages. 17 × 24.5 cm. Paper. Available free to developing countries, but restricted distribution to developed countries.

Directories of Germplasm Collection

This description applies to all the books in the following section: These directories list germplasm holdings in institutes around the world. The information contained therein has been provided by the curators. These data are a preliminary atempt to assess total holdings and the information aids scientists in making contact with other scientists working on the same crop.

G.A. Juvik, R.L. Bernard and H.E. Kauffman
Directory of Germplasm Collections; 1.II Food Legumes (Soyabean)

1985. 53 pages. 22 × 29.7 cm. Paper. Published in cooperation with the University of Illinois, Urbana-Champaign, USA. Available free to developing countries, but restricted distribution to developed countries.

T. Lawrence, J. Toll and D.H. van Sloten

Directory of Germplasm Collections; 2. Root and Tuber Crops

1986. 178 pages. 21 × 29.7 cm. Paper. Available free to developing countries, but restricted distribution to developed countries.

Conferences and Meetings

IBPGR Advisory Committee on *In Vitro* Storage: Report of the Second Meeting

1985. 15 pages. 21 × 29.7 cm. Paper. Available free to developing countries, but restricted distribution to developed countries.

This document reports the findings of the second meeting of one of the IBPGR's two Standing Committees on Conservation.

ICARDA

The International Center for Agricultural Research in the Dry Areas

P.O. Box 5466, Aleppo, Syria
Cable: ICARDA - Aleppo
Telex: 331206 SY; 331208 SY; 331263 SY
Telephone: 550465; 551280; 213433 Tel Hadya;
213477 Tel Hadya

General Information

Objectives:

To improve the agricultural systems and major food crops of the drier regions of Western Asia and North Africa. Mediterranean-type of climate of cool, moist winter and hot, dry summers, and the high elevation plateaux, with extremes of winter cold and summer heat, and snow cover for up to five months a year.

Research Highlights

ICARDA

Research Highlights 1985

1986. ___ pages. 16.8 × 23 cm. ISSN 0255-643-X. Paperback.

This brochure describes the highlights of ICARDA's research during 1985. The topics highlighted give a partial insight into the continuing work of ICARDA scientists and the progress they have already made.

Annual Reports

ICARDA

Annual Report 1985

1986. 378 pages. 19.8 × 23.8 cm. ISSN 0254-8313. Paperback.

This annual report gives an overview of the center's research activities and results in 1985 on ICARDA's mandated crops. Activities of the genetics resources, computer services, communication and documentation, visitors' services, and collaborative projects with advanced institutions are also reviewed and lists of scientific publications and seminars given during the year are included.

Conference, Workshop, and Symposia Proceedings

S. Miladi, S. Mahgoub and P. Neate (Editors)

Interfaces between Agriculture, Food Science, and Nutrition: Proceedings of a National Workshop in Sudan, 10-15 Dec 1983

1985. 211 pages. Paperback.

This book is the outcome of this workshop. It comprises papers presented by agriculturalists, economists, food scientists, sociologists, and nutritionists, participants in the workshop, and their recommendations to foster future cooperation among them.

T.L. Nordblom, A.K.H. Ahmed, and G.R. Potts (Editors)

Research Methodology for Livestock On-Farm Trials: Proceedings of a Workshop held at Aleppo, Syria, 25-28 Mar 1985

1985. 313 pages. Paperback. ISBN 0-88936-446-X. Publisher: IDRC and ICARDA.

This book contains 12 studies describing methods used in conducting livestock on-farm trials, primarily in the Middle East and Africa regions. Also included arc five methodological summaries covering definition of research problems, trial design, farmer participation, criteria for evaluation, and future directions for lofts.

J.P. Srivastava and L.T. Simarski (Editors)

Seed Production Technology

1986. 287 pages. 17.1 $\times$ 24 cm. Paperback.

This book is a compilation of the lectures delivered by specialists to the trainees participating in the Third Seed Production Course, 1984.

Miscellaneous Reports and Information Brochures

Harvest of Research: Highlights of the IFAD/ICARDA Nile Valley Project 1979-1985

1985. 48 pages. Soft cover.

This publication tells the story of the Nile Valley Project between 1979 and 1985 to improve faba bean production in Egypt and Sudan and the achievements of the project.

B.H. Somaroo and Y.J. Adham (Editors)

Barley Germplasm Catalog I

1986. 413 pages. Soft cover.

The catalog intended as an aid for researchers lists 8,000 barley germplasm accessions with information on their country of origin and the evaluation data on 22 different agro-morphological, physiological and other traits.

ICARDA: A Partner in Cereal Improvement

1985. 74 pages. 16.5 × 24 cm. Soft cover.

This publication acquaints the readers with the kinds of services ICARDA, and in particular its cereal improvement program, offers to national cereal programs and some of the joint accomplishments emerging from their work together.

Technical Manual

P. Williams, F. Jaby El-Haramein, H. Nakkoul and S. Rihawi (Editors)

1986. 142 pages. 17 × 24 cm. Paperback.

This manual provides instructions to breeders and staff involved in quality screening of cereals, pulses, and forage crops. It gives a brief description of the tests used, their principles, and interpretation, and includes instructions for each test.

Periodicals

RACHIS Newsletter (Barley and Wheat Newsletter)

July 1985. 64 pages. Vol. 4 (2).

This newsletter is published biannually. It contains scientific articles, short communications, book reviews, news items about training, conferences and scientists in barley and wheat. Arabic versions are also published with the financial support of IDRC, Ottawa, Canada.

FABIS Newsletter (Faba Bean Information Service), supported by IDRC, Ottawa, Canada

December 1985. 43 pages. No. 13.

This newsletter is published triannually. It contains short scientific research articles, review articles on specific areas of faba bean research, book reviews, announcements, and news items about training and conferences on faba bean.

LENS Newsletter (Lentil Experimental News Service), supported by IDRC

1985. 54 pages. Vol. 12 (2).

This newsletter is published biannually in cooperation with the University of Saskatchewan, Canada. It contains review articles on specific areas of lentil research, news items about conferences on lentils, and descriptions of ICARDA publications.

ICRISAT

International Crops Research Institute for the Semi-Arid Tropics

Patancheru P.O. Andhra Pradesh 502324, India
Cable: ICRISAT, Hyderabad - Telex: 0152-203

General Information

The International Crops Research Institute for the Semi-Arid Tropics (ICRISAT) was founded in 1972 and is one of 13 centers in an international research network operating under the aegis of the Consultative Group on International Agricultural Research (CGIAR). Its headquarters and main research farm, ICRISAT Center, is near the village of Patancheru, 25 km northwest of Hyderabad, India. Other ICRISAT scientists are posted in several countries of Africa, in Mexico, and in Syria.

The mandate of ICRISAT is to:
1. Serve as a world center for the improvement of grain yield and quality of sorghum, millet, chickpea, pigeonpea, and groundnut, and to act as a world repository for the genetic resources of these crops.
2. Develop improved farming systems that will help to increase and stabilize agricultural production through more effective use of natural and human resources in the seasonally dry semi-arid tropics.
3. Identify constraints to agricultural development in the semi-arid tropics and evaluate means of alleviating them through technological and institutional changes.
4. Assist in the development and transfer of technology to the farmer through cooperation with national and regional programs, and by sponsoring workshops and conferences, operating training programs, and assisting extension activities.

Since its foundation ICRISAT has realized the importance of communicating its research findings and production technology via a wide range of publications. The ones listed here are bound in paper covers unless otherwise stated.

Ordering Instructions

Per-copy prices are listed separately for:
- highly-developed countries (HDCs), expressed in U.S. dollars;
- less-developed countries (LDCs), also expressed in U.S. dollars;
- India and other countries in the Indian subcontinent (India), expressed in rupees at a rate equivalent to the LDC price.

The LDC and India prices are quoted at cost, to permit maximal distribution among clients in national agricultural research and extension programs in developing countries.

Book trade discounts are available on request. Other orders for five or more copies are discounted by 30%.

Prepayment is required, based on a proforma invoice. Payment is accepted in Indian rupees, in U.S. dollars or negotiable hard currencies, or in UNESCO coupons.

Prices, and cost of postage and packing (P & P), are quoted for all priced publications and must be remitted in full. Out-of-print publications are available as microfiche copies, at flat-rate page charges.

Organizations that formally exchange publications with ICRISAT Library, journal editors who review ICRISAT publications, and national program staff who collaborate with ICRISAT research programs are entitled to receive single free copies.

Publications will be dispatched via air bookpost unless sea mail is specifically requested. Experience shows that surface mail packages are often delayed for months, do not always reach their destinations, and may be damaged.

Send orders to:

> Information Services
> ICRISAT
> Patancheru
> Andhra Pradesh 502 324
> India

Research Highlights

ICRISAT Research Highlights 1985

1986. 48 pages. 17 × 23.5 cm. Saddle stitched. ISSN 0257-2532. Free.

Reports advances in ICRISAT's research work during 1985 on its five mandate crops: sorghum, pearl millet, chickpea, pigeonpea, and groundnut. Illustrated in color, the text focuses on the Institute's multidisciplinary approach to problem-solving in the development of agriculture in the semi-arid tropics.

ICRISAT: Progrés de la recherche, 1985.

1986. 48 pages. 17 × 23.5 cm. Saddle stitched. ISSN 0257-2494. Free.

Fait le point sur les recherches de l'ICRISAT au cours de l'année 1985 sur les cinq cultures de son mandat: sorgho, mil, pois chiche, pois d'Angole et arachide. Illustre en couleurs, le texte est axe sur l'approche multidisciplinaire de l'Institut à la solution des problèmes relevant du développement de l'agriculture dans les zones tropicales semi-arides.

Conferences

Proceedings of the International Sorghum Entomology Workshop, 15-21 July 1984.

1985. 434 pages. 17 × 23.5 cm. Paperback. ISBN 92-9066-108-9. HDC $26.40, LDC $8.80 plus airmail ($13.10) postage.

Presents the proceedings of the workshop held at College Station, Texas, USA. Participants discussed the need for sorghum entomology research the world over, with special reference to the developing world. Reports are given on sorghum pests in eastern and West Africa, Southeast Asia, India, Australia, and most of the Americas. Other sections concern specific pests and a final section is devoted to breeding for insect resistance in sorghum and other control approaches.

Summary Proceedings of the Consultative Group Meeting on Collaborative Research on Groundnut Rosette Virus, 13-14 April 1985.

1985. 28 pages. 17 × 23.5 cm. Saddle stitched. ISBN 92-9066-113-5. HDC $3.00, LDC $1.00 plus airmail ($2.20) postage.

Contains the summaries of eight papers presented at the meeting, which brought together representatives of various research groups working on groundnut rosette virus disease from all over the world to discuss recent findings and plans for future research. Recommendations for a coordinated international collaborative plan and the guidelines set down for each research group are also included.

Workshop on the Training Needs for Dryland Agriculture in India, 17-18 July 1985.

1985. 48 pages. 21.5 × 28 cm. Saddle stitched. ISBN 92-9066-114-3. HDC $3.00, LDC $1.00 plus airmail ($3.70) postage.

Reports on this 2-day workshop held with particular reference to deep Vertisol technology. The recommendations include suggestions for effective watershed management and development, stronger infrastructural facilities and staff for training, and measures to strengthen research resources.

Summary Proceedings of the Working Group on Agroforestry Research in the Semi-Arid Tropics, 5-6 August 1985.

1986. 68 pages. 17 × 23.5 cm. Saddle stitched. ISBN 92-9066-116-X. HDC $4.20, LDC $1.40 plus airmail ($2.80) postage.

This meeting brought together participants from industrial, research, university, donor, and international organizations, to share ideas, methodologies, and results in agroforestry research. Recommendations were made for both bioscientific and socioeconomic research, and for developing collaborative linkages.

Farming Systems Principles for Improved Food Production and the Control of Soil Degradation in the Arid, Semi-Arid, and Humid Tropics, 20-23 June 1983.

1986. 36 pages. 17 × 23.5 cm. Saddle stitched. ISBN 92-9066-115-1. HDC $3.00, LDC $1.00 plus airmail ($2.00) postage.

Comprises the summary proceedings of an experts' meeting sponsored by the UN Environment Programme, the main objective of which was to identify areas of farming systems research and development having potential to significantly increase food production and promote soil conservation for selected agroecological zones. Guidelines and research priorities are given for farming systems to prevent soil degradation in arid lands, the semi-arid tropics, and the humid tropics.

Miscellaneous

ICRISAT in Africa

1986. 60 pages. 17 × 23.5 cm. Saddle stitched. Free.

Describes the collaborative work of ICRISAT scientists in semi-arid tropical countries in Africa. Illustrated in color.

L'ICRISAT en Afrique

1986. 60 pages. 17 × 23.5 cm. Saddle stitched. Free.

Décrit les travaux collaboratifs des chercheurs de l'ICRISAT dans les pays tropicaux semi-arides en Afrique. Illustré en couleurs.

Bibliographies

Sorghum and Millet Information Center
Sorghum bibliography, 1982

1986. 396 pages. 21.5 × 28 cm. Paperback. ISBN 92-9066-117-8. Free. Airmail postage, $12.00.

Contains 1616 annotated entries, grouped under broad headings such as Agronomy, Pathology, Entomology, etc. Detailed subject and author indexes are included.

Sorghum and Millet Information Center
Millet bibliography, 1982.

1985. 368 pages. 21.5 × 28 cm. Paperback. ISBN 92-9066-110-0. Free. Airmail postage, $12.00.

Contains 991 annotated entries, grouped in three major categories - pearl millet, minor and other millets, and wild relatives. The bibliography includes author and subject indexes.

IFPRI

International Food Policy Research Institute

1776 Massachusetts Avenue, N.W., Washington, D.C. 20036, USA
Cable: IFPRI - Telex: 440954
Telephone: (202) 862-5600

General Information

Objectives

Research at IFPRI is aimed at contributing to the reduction of hunger and malnutrition in low-income countries through the analysis of underlying processes that extend beyond the narrowly defined food sector. The results of IFPRI's research and analysis are intended to assist policymakers in developing countries make informed decisions on food policy issues. IFPRI works closely with other international and developing nation institutions to understand food production, consumption, and trade processes; to develop new options for policymakers; and to assess the efficacy of existing policies. Individual copies of most IFPRI publications are free but multiple copies will carry a fee plus shipping costs. The fee varies with size and weight of publications.

Annual Report

IFPRI Report 1985

May 1986. 63 pages. 17.8 × 25.4 cm. Saddle stitched.

The annual report summarizes the year's work in each of the Institute's five main programs — food data evaluation, food production policy, agricultural growth linkages, food consumption and nutrition policy, and international food trade and food security — as well as the year's seminars and collaborative work. *IFPRI Report 1984* is also available.

Monographs

Research Reports

No. 50
Victor J. Elias

Government Expenditures on Agriculture and Agricultural Growth in Latin America

October 1985. 68 pages. 17.8 × 25.4 cm. Saddle stitched. ISBN 0-89629-051-4.

This study evaluates how government expenditures on agriculture affected agricultural output between 1950 and 1980 in Argentina, Bolivia, Brazil, Chile, Colombia, Costa Rica, Mexico, Peru, and Venezuela.

No. 51
Michel Petit

Determinants of Agricultural Policies in the United States and the European Community

November 1985. 80 pages. ISBN 0-89629-052-2.

This report presents a conceptual framework for analyzing why agricultural policies are what they are, how and under which forces they evolve, and how much of their future evolution can be predicted. Three case studies involving U.S. and European agricultural and agricultural trade policies illustrate and specify the framework.

No. 52
Leonardo Paulino

Food in the Third World: Past Trends and Projections to 2000

June 1986. 76 pages. ISBN 0-89629-054-9.

The trends In production, consumption, and trade of basic food staples (cereals, roots and tubers, pulses, groundnuts, and bananas and plantains) in 105 developing countries are examined and projections developed based on output and income trends.

No. 53
Ulrich Koester

Regional Cooperation to Improve Food Security Among the Southern and Eastern African Countries

____________ 1986. ____ pages. ISBN 0-89629-053-0.

This report investigates whether regional cooperation can improve food security for developing countries. It identifies the determinants of successful cooperation schemes, and it develops a methodology for quantifying the potential benefits. Specifically it examines the potential for food security in the nine African countries comprising SADCC.

No. 54
Padma Desai

Weather and Soviet Grain Yields

____________ 1986. ____ pages. ISBN____________.

The influence of weather on Soviet grain yields is measured and the role of systemic and policy factors in weather-adjusted yields is assessed. Methodologies are devised for analyzing such issues.

No. 55
T. Ademola Oyejide

Effects of Trade Regime and Exchange Rate Policy on the Structure of Incentives to Agriculture in Nigeria

__________ 1986. ____ pages. ISBN__________.

This report measures, in terms of relative prices, the incidence of protection accorded by trade and exchange rate policies to agriculture in relation to other sectors of the economy; assesses how trade and exchange rate policies affect intra- and intersectoral resource allocation; and examines how the oil sector affected production incentives in agriculture.

Books

John W. Mellor and Gunvant M. Desai, eds.

Agricultural Change and Rural Poverty: Variations on a Theme by Dharm Narain

1985. 233 pages. 15.8 × 23.5 cm. Hardcover. ISBN 0-8018-3275-6. US$24.95. The John Hopkins University Press and the Oxford University Press, for IFPRI.

Leading development specialists examine the complexities of the issue of whether the green revolution has left the poor behind. Conclusions with major implications for development policy are presented, and the experiences of and lessons from India, Japan, Southeast Asia, and Africa are compared.

Order from publisher:
        The John Hopkins University Press
        701 W. 40th Street
        Suite 275
        Baltimore, MD 21211, USA

In India, the publisher is:
        Oxford University Press
        YMCA Library Building
        Jai Singh Road
        Post Box No. 43
        New Delhi 110001, INDIA.

Peter Hazell, Carlos Pomareda, and Alberto Valdes, eds.

Crop Insurance for Agricultural Development: Issues and Experience

1986. 322 pages. 15.8 × 23.5 cm. Hardcover. ISBN 0-8018-2673-X. US$32.50. The Johns Hopkins University Press for IFPRI.

Crop Insurance is assessed by presenting the economic theory behind it, testing it against empirical data from several countries, and putting it into perspective with other policy options. The book suggests how the efficiency of crop insurance programs could be increased and possible alternative policies.

Order from JHU Press (see under previous publication for address).

Conference Proceedings

IFPRI and the German Foundation for International Development (DSE).

Summary Proceedings of the Workshop on Cereal Yield Variability

1986. ____ pages. ISBN 0-89629-306-8.

These papers deal empirically with a range of issues associated with cereal yield variability. The nature of changing patterns of yield variability for different cereals in various regions is documented, and alternative approaches to reducing variability are examined. The relationship between changes in yield variability and yield correlations and such causal factors as changes in agricultural technologies, weather, irrigation, input availability, and farming systems is analyzed.

Miscellaneous Reports

Ojetunji Aboyade

Administering Food Producer Prices in Africa: Lessons from International Experiences

December 1985. 58 pages. 21.59 × 27.30 cm. Saddle stitched. ISBN 0-89629-304-1.

This work explores the possible effects of increasing price incentives to farmers in Sub-Saharan Africa. The institutional factors and administrative arrangements necessary for a viable producer-price incentive system are examined against the background of case-study experiences in Africa and Asia.

Harold Alderman

The Effect of Food Price and Income Changes on the Acquisition of Food by Low-Income Households

May 1986. 100 pages. 21.59 × 27.30 cm. Saddle stitched. ISBN 0-89629-305-X.

This paper reviews the methodological approaches used and the empirical results obtained in quantifying the extent to which the poor are more responsive to price changes than wealthier consumers. It identifies generalizable findings that may be used to approximate price responses and provides insights into the adjustments low-income households make in food consumption in response to price changes.

IITA

International Institute of Tropical Agriculture

P.M.B. 5320, Ibadan, Nigeria
Cable: TROFOUND, IJEKA, NIGERIA
Telex: TROPIB NG 31417

Publications of the International Institute of Tropical Agriculture 1984

About IITA

The International Institute of Tropical Agriculture is a non-profit research and training institute established in 1967 under the Laws of the Federal Republic of Nigeria. Like its other 12 sister institutions, IITA is principally financed through the Consultative Group on International Agricultural Research which is an informal group of donor countries, development banks, foundations, and agencies.

The mandate

IITA is responsible for developing viable alternatives to shifting cultivation in the humid and sub-humid tropics. Also, the Institute has a global mandate for research to improve cowpeas, yams, and sweet potatoes as well as carrying out research in Africa on maize, rice, cassava, and soybean. It organizes training, conferences and workshops to increase the number of well qualified persons to carry out research and development.

Books published by commercial companies should be obtained through the book stores.

Grain Legume Improvement Program, IITA
Large Scale Soybean Production in Nigeria

1985. 10 pages. 21 × 27 cm. Hard cover. HDC $3.00, LDC $3.00 plus airmail ($6.30) or surface mail ($4.20) postage.

M. Ashraf, P. Balogun and A. Jibrin
A Case Study of On-Farm Adaptive Research in the Bida Agricultural Development Project in Nigeria

1985. 32 pages. 21 × 27 cm. Hard cover. HDC $3.00, LDC $3.00 plus airmail ($8.50) or surface mail ($4.20) postage.

This publication is the first in a series of IITA On-Farm Bulletins. It describes work undertaken with the Agronomy Section of the Bida Agricultural Development Project in Niger State, Nigeria.

M.C. Palada, W.O. Vogel and H.J.W. Mutsaers

Report on Exploratory Survey of Ijaiye-Imini Pilot Research Area, Oyo State, Nigeria

1985. 59 pages. 21 × 27 cm. Hard cover. HDC $3.00, LDC $3.00 plus airmail ($11.80) or surface mail ($5.20) postage.

The result of an exploratory survey carried out by a multidisciplinary team in preparation of on-farm technology testing. It describes the farming environment and the farming systems in the area and the constraints and opportunities observed by the team.

H.H.W. Matsaers

On-farm Research Training Workshops

1985. 40 pages. 21 × 27 cm. Hard cover. LDC $3.00, LDC $3.00 plus airmail ($9.60) or surface mail ($4.50) postage.

This publication is written as a set of practical hints for an efficient organization of On-farm Research (OFR) training workshops, for use by IITA trainers.

H.W. Rossel and G. Thottappilly

Virus Diseases of Important Food Crops in Tropical Africa

1985. 61 pages. 15 × 22 cm. Hard cover. HDC $5.00, LDC $5.00 plus airmail ($11.60) or surface mail ($7.20) postage.

This book contains information about the geographical distribution, symptoms, identification and control of the most prevalent virus diseases of some of the continent's principal staple food crops.

B.B. Singh, S.R. Singh, L.E.N. Jackai and S.A. Shoyinka

General Guide for Cowpea Cultivation and Seed Production

1986. 16 pages. 21 × 28 cm. Hard cover. HDC $3.00, LDC $3.00 plus airmail ($7.95) or surface mail ($4.20) postage.

The publication provides information on cowpea culture to serve as a guide for farmers and seed producing agencies.

S.K. Hahn

Les Plantes a Racines et Tubercules Tropicales Amelioration et Utilisation

1986. 32 pages. 21 × 27 cm. Hard cover. HDC $3.00, LDC $3.00 plus airmail ($8.50) or surface mail ($4.20) postage.

D'apres une communication presentee lors d'une conference organisee par le Bureau agricole du Commonwealth, et intitulee: "Accroissement de la production agricole en Afrique", 13-17 Fevrier 1984, Arusha, Tanzanie.

Efren Oro

Basic Course in Photography

1986. 39 pages. 14 × 22 cm. Soft cover. HDC $3.00, LDC $3.00 plus airmail ($6.30) or surface mail ($4.20) postage.

IITA

Special Issue Tropical Grain Legume Bulletin — No. 32

1986. 173 pages. 21 × 27 cm. Hard cover. HDC $5.00, LDC $5.00 plus airmail ($38.00) or surface mail ($10.20) postage.

Based on papers presented at the first World Cowpea Research Conference, held at IITA, Ibadan, Nigeria - 5 to 9 November, 1984. These papers contained valuable information. To ensure that the salient features of the papers are documented, the Grain Legume Improvement Program at IITA decided to publish them as a Special Supplement of the Tropical Grain Legume Bulletin.

ILCA

International Livestock Centre for Africa

Address: P.O. Box 5689   Addis Ababa
Cable Address: ILCAF ADDIS ABABA
Telex Address: ILCA 21207 Telephone: 18-32-15

ILCA

General Information

Mandate: To assist national efforts which aim to effect a change in production and marketing systems in tropical Africa so as to increase the sustained yield and output of livestock products and improve the quality of the life of the people of this region.

Centre Objectives

ILCA is an international centre of livestock research, training and documentation. Through research, it seeks to increase the production and sale of livestock and livestock products. This research effort is supported by an extensive documentation service and by the provision of training programmes for national research workers. ILCA's goal of increasing livestock production originates in the premise that the output of the arable and pastoral land of Africa is the major determinant of economic growth, and that inadequate increase in livestock and crop output in most African countries in recent years are the key reason for the unsatisfactory rate of economic growth in these countries. The role of livestock in stimulating economic growth derives from their capacity to generate rapid increases in the cash income of subsistence farmers, enabling these farmers to purchase the agricultural inputs needed to generate large increase in food grain production. In the traditional agricultural systems of Africa, there is a strong complementarity between livestock output and crop production, with increase in the former quickly leading to an upturn in the latter. It is this relationship which ILCA specifically seeks to exploit.

Annual Reports

ILCA

ILCA Annual Report

Annually since 1981; latest 1985. Varies (approx. 80 pages). 28 × 21 cm. Saddle stitched. ISSN 0255-0040 (English) and 0255-3473 (French).

Highlights the work of ILCA's field research programmes in Ethiopia, Nigeria, Mali, Kenya, and Botswana as well as work on livestock policy, trypanotolerance, forage legume agronomy, small ruminants and camels, nutrition, aerial survey, computer services, library and documentation, training and publications.

Available in English and French.

Conferences

R.T. Wilson and D. Bourzat

Small Ruminants in African Agriculture
Les petits ruminants dans l'agriculture Africaine

November 1985. IX, 261 pages. 21 × 29.7 cm. Paperback. ISBN 92-9053-069-3. $30.00. Airmail postage included.

The 24 papers in these proceedings deal with goat and sheep research in 14 African countries. The role of these animals in a wide range of ecological zones and agricultural systems is described as well as their reproductive and productive performance, nutrition and mortality and morbidity.

Bilingual (English and French).

N. de Ridder, H van Keulen, N.E. Seligman, and P.J.H. Neate (editors)

Modelling of Extensive Livestock Production Systems

January 1986. 355 pages. 21 × 29.7 cm. Adhesive binding, hot melt-paperback. ISBN 92-9053-073-1. $40.00. Airmail postage included.

The proceedings of the ILCA/ARO/CABO workshop held at Bet Dagan, Israel, in February 1986 focus on four main aspects of modelling extensive livestock production systems: primary production, factors affecting feed intake and its prediction, secondary production, and management and economics. The progress achieved in this work at the three centres, as well as desirable lines of development in the future, are discussed.

R. von Kaufmann, S. Chater, and R. Blench

Livestock Systems Research in Nigeria's Subhumid Zone

March 1986. 497 pages. 21 × 29.7 cm. Adhesive binding, hot melt-paperback. ISBN 92-9053-074-X. $56.00. Airmail postage included.

Based on the Second ILCA/NAPRI Symposium held in Kaduna, Nigeria, in 1984, this collection of papers presents the results of livestock systems research in Nigeria's subhumid zone. The authors of the selected papers give basic information about the systems studied and discuss the interactions between crop and livestock production, as well as the reactions of the Fulani agropastoralists participating in the research and the role of national research institutes and extension services in ILCA's present and future work.

ILCA Research Report

No. 12
K. Agyamang and L.P. Nkhonjera

Evaluation of the Productivity of Crossbred Dairy Cattle on Smallholder and Government Farms in the Republic of Malawi

February 1986. 47 pages. 17.5 × 25.5 cm. Adhesive binding, hot melt-paperback. ISBN 92-9053-072-3. $8.00 Airmail postage included.

The emphasis in this ILCA research report is on the evaluation of reproductive and productive performance of crossbred cows on smallholder farms in the Southern Region of Malawi. The performance of crossbred cows on three research stations, as well as that of the pure-Friesian herd at Mikolongwe, is also evaluated.

No. 13
K.T. Wagenaar and A. Diallo

Productivity of Transhumant Fulani Cattle in the Inner Niger Delta of Mali

October 1986. 55 pages. 17.5 × 25.5 cm. Adhesive binding, hot melt-paperback. ISBN 92-9053-079-0. $9.00. Airmail postage included.

The thirteenth report in the series of ILCA Research Reports deals with a range of performance traits and the productivity of transhumant Sudanese Fulani Zebu cattle in the inner Niger delta of Mali. The analyses carried out are based on data collected on 2550 animals in three herds studied over the 1978-83 period. The results of the study are compared with findings reported in the literature on Sudanese Fulani cattle in Mali, the related White Fulani in Nigeria and the Gobra in Senegal.

Manuals

M.J. Nicholson and M.H. Butterworth

A Guide to Condition Scoring of Zebu Cattle

November 1985. 30 pages. 15 × 21 cm. Saddle stitched. Paperback. ISBN 92-9053-068-5. $4.00. Airmail postage included.

The authors of this handy manual give background information on condition scoring and describe a 9-score method suitable for assessing body condition of *Bos indicus* (Zebu) cattle in Africa. The various conditions observed are shown in black-and-white photographs presenting side and rear views of the scored animals. The manual also gives information on the statistical analysis of condition scoring data.

Abiye Astatke, S. Bunning, and F.M. Anderson

Building Ponds with Animal Power in the Ethiopian Highlands

January 1986. 44 pages. 14.8 × 21 cm. Saddle stitched. Paperback. ISBN 92-9053-070-7. $11.00. Airmail postage included.

Many people throughout the Third World lack access to adequate water supplies for household consumption or for livestock. This problem can be alleviated by building ponds to trap surface runoff during the rainy season. This manual on building ponds in Ethiopia provides useful information and guidance to those involved in pond excavation using animal power.

Audiovisual Materials

ILCA

ILCA's Highlands Programme

1985. 46 slides (1-projector version). 13 minutes. $25.00 Airmail postage included.

This tape/slide presentation describes the work of ILCA's Highlands Programme, whose staff work in the highlands of Ethiopia. Research activities at the Programme's two main research stations, Debre Zeit and Debre Berhan, are described. These include improvements in cultivation tools, excavation of surface ponds, use of biogas digesters, and the integration of forage legumes into the traditional farming systems.

Available in English only.

ILCA

Research at ILCA's Debre Zeit Station

1985. 28 slides (1-projector version). 8 minutes. $20.00. Airmail postage included.

This tape/slide presentation describes the research work of ILCA's Highlands Programme at the Debre Zeit research station, 50 km southeast of Addis Ababa in the Ethiopian highlands. The work described includes improved techniques for managing Vertisols, improved cultivation implements, the use of biogas digesters, and forage legumes.

Available in English only.

ILCA

Research at ILCA's Debre Berhan Station

1985. 38 slides (1-projector version). 10 minutes. $20.00. Airmail postage included.

This tape/slide presentation describes the work of ILCA's Highlands Programme at the Debre Berhan research station, 120 km northeast of Addis Ababa in the Ethiopian highlands. The work described includes the excavation of surface water ponds using animal power, improvements in dairy processing techniques, the improved use of sheep's wool and the introduction of forage legumes.

Available in English only.

ILCA

The Trypanotolerance Network

1986. 87 slides. (1-projector version). 25 minutes. $45.00. Airmail postage included.

This tape/slide presentation describes the problem of trypanosomiasis in tsetse-infested Africa, the research network established and coordinated by ILCA Nairobi, the aims and operation of the network, network training and computer analysis. The data collection work in 10 African countries is also described and some recent results are presented.

Available in English only.

Audiovisual Information Packages

The following audiovisual information packages produced by ILCA may be purchased subject to availability. Each package includes 35mm slides, cassette with sync pulse and script.

ILCA

The International Livestock Centre for Africa (ILCA)

1986. 88 slides (3-projector version); 71 slides (1-projector version). 19 minutes. $60.00 and $35.00, respectively. Airmail postage included.

This tape/slide presentation describes the problem of low food production in Africa, the role that livestok can play in stimulating increases in production, the mandate and location of ILCA, the work of ILCA's zonal research programmes and central research units, ILCA's training and information activities, and the sources of ILCA's funding.

Available in English and French.

N.B Versions of the 3-projector version of the ILCA tape/slide presentation are now available on UMATIC and VHS video cassettes for use with both PAL and NTSC systems. Prices available on request.

ILCA
Livestock in Niger

1986. 52 slides (1-projector version). 11 minutes. $25.00. Airmail postage included.

This tape/slide presentation describes the importance of livestock and their production in Niger. It summarises current research approaches to livestock improvements and discusses future prospects.

Available in French only.

Bibliographies, Directories, Abstracts

I. Haque, Desta Beyene, and Marcos Sahlu
Bibliography on Soils, Fertilizers, Plant Nutrition and General Agronomy in Ethiopia

1985. 62 pages. 17 × 24 cm. Adhesive binding, hot melt- paperback. ISBN 92-9053-006-9. $15.00. Airmail postage included.

This is the first comprehensive bibliography of research in Ethiopia on soils, fertilizers, plant nutrition and general agronomy. It is intended for use by researchers, development agents, planners and administrators.

M.H. Butterworth, Michael Hailu, Marcos Sahlu, and Sirak Teklu
Beef Cattle Production from Tropical Pastures: A Descriptive Bibliography

November 1985. 118 pages. 17 × 24 cm. Adhesive binding, hot melt-paperback. ISBN 92-9053-057-X. $20.00. Airmail postage included.

This bibliography lists references to research work conducted on beef cattle production under different management systems in the tropics, with special reference to methods by which productivity may be increased. The references were drawn from *Beef cattle nutrition and tropical pastures* (Longmans, 1985) and from ILCA's documentation data base.

ILCA/IDRC team in Botswana
Index to Livestock Literature Microfiched in Botswana

December 1985. 154 pages. 17 × 24 cm. Adhesive binding, hot melt-paperback. ISBN 92-9053-063-4. Airmail postage included.

Non-conventional literature available in Botswana on the country's livestock industry was first screened by ILCA's documentation team in 1978 and listed in the ILCA/Botswana Working Document No. 4. The present index includes an additional 731 documents microfiched in Botswana in 1983. All listed references are available on-line from ILCA's data base.

J.E. Sumberg

Gliricidia sepium(Jacq.) Steud: A Selected Bibliography

January 1986. 18 pages. 17 × 24 cm. Saddle stitched. Paperback. ISBN 92-9053-061-8. $4.00. Airmail postage included.

References to research work conducted on *Gliricidia* over the last decade are scattered in journals of several disciplines. This bibliography provides quick access to the biology, use and management of *Gliricidia* in the humid and subhumid tropics, with particular reference to its potential in intensified smallholder farming systems.

Tesfai Berhane and Negussie Akalework

Index to Livestock Literature Microfiched in Zimbabwe

February 1986. 243 pages. 17 × 24 cm. Adhesive binding, hot melt-paperback. ISBN 92-9053-064-2.

This index comprises 397 non-conventional documents selected and microfiched by ILCA's documentation team in Zimbabwe in 1981 and a further 1145 documents microfiched during the follow-up mission in 1983. The references listed in the index have been incorporated into the ILCA data base and are accessible on-line.

Negussie Akalework et Mekonen Assefa

Index des Documents Microfiches au Benin

April 1986. 73 pages. 17 × 24 cm. Adhesive binding. ISBN 92-9053-077-4.

This index comprises references to non-conventional literature microfiched by ILCA's documentation team in Benin in 1985. The references have been incorporated into the ILCA data base and are available on-line.

ILRAD

International Laboratory for Research on Animal Diseases

ILRAD

P.O. Box 30709, Nairobi, Kenya
Cable: ILRAD Nairobi, Kenya
Telex: 22040 ILRAD
Telephone: Nairobi 592311

General Information

Objectives:

The International Laboratory for Research on Animal Diseases (ILRAD) was established in 1973. The laboratory complex was constructed on a 70-hectare site donated by the Kenya Government at Kabete, on the outskirts of Nairobi, Kenya.

ILRAD's mandate is to conduct basic research leading towards the development of safe, effective and economically feasible measures to control livestock diseases which seriously limit world food production. Emphasis is on disease control by immunological means. However, this does not preclude research on other promising disease control measures, such as the strategic use of chemotherapy, genetic improvements and vector control.

ILRAD also carries out training designed to support the development of scientific and field personnel who can extend research on pressing animal health problems - primarily in Africa, but also in other parts of the world. Scientists and technicians are trained to carry out the most effective disease control programs possible with the means currently available. At the same time, they are prepared to initiate new field programs when improved control methods are ready for introduction.

Major research programs:

Two important livestock diseases caused by protozoan parasites were chosen as the first targets for research at ILRAD: East Coast fever, which is a virulent form of theileriosis, and African animal trypanosomiasis.

Annual Reports

ILRAD
ILRAD 1985

1986. 76 pages. 17.5 X 25 cm. Saddle stitched. ISBN 92-9055-085-6. Free.

ILRAD's 1985 annual report summarizes research activities and results aimed at improving control measures for African animal trypanosomiasis and East Coast fever, a virulent form of theileriosis. Training and information activities are summarized and scientific articles and other publications during the year are listed.

Published in English and French.

Conferences

A.D. Irvin (Editor)

Immunization against East Coast Fever

1986. 32 pages. 21 × 29 cm. Saddle stitched. ISBN 92-9055-286-7. Free.

Immunization against East Coast Fever presents the proceedings of a workshop held at ILRAD in September 1985 to discuss the most appropriate ways of collecting, handling and analysing data on the performance and productivity of cattle immunized against East Coast fever by 'infection and treatment'. This workshop was a sequel to one held in 1984 in which the different methods of immunization against East Coast fever were discussed.

A.D. Irvin (Editor)

Immunization against Theileriosis in Africa

1985. 167 pages. 21 × 29 cm. Saddle stitched. ISBN 92-9055-258-9. Free.

Immunization against Theileriosis in Africa presents the proceedings of a joint workshop sponsored by ILRAD and FAO in Nairobi in October 1984. Participants from 11 African countries affected by theileriosis met with invited specialists to discuss the extension of 'infection and treatment' immunization against East Coast fever. The prospects of similarly controlling related sub-Saharan theilerioses were also considered.

Miscellaneous

ILRAD
W.I. Morrison (Editor)

The Ruminant Immune System in Health and Disease

1986. 580 pages. 16 × 23.5 cm. Saddle stitched. ISBN 0-521-3244-2. Free.

This volume reviews the immune system of domestic ruminants, with particular emphasis on mechanisms of immunity and resistance to infectious disease. Topics of comparative interest in other species are also covered.

The cellular components of the immune system are discussed first, including ontogeny, structural organization, and how they interact during the generation of immune responses. This is followed by information on specific aspects of humoral and cell-mediated immune responses, and finally a section on the relevance of various types of immune responses in protection against a number of important infectious diseases of ruminants.

ILRAD

Scientific Publications from ILRAD: 1984-85 Update

1986. 12 pages. 10.5 × 21.5 cm. Saddle stitched. Free.

Scientific publications from ILRAD is a brochure listing books, chapters and articles in international journals published by ILRAD staff members. A new edition is published, in English, every 2 years.

ILRAD

Training Opportunities at ILRAD

1985 (French edition 1986). 10 pages. 10.5 × 21.5 cm. Concertina. Free.

Training opportunities at ILRAD gives a brief description of ILRAD's facilities and research programs. Courses and training programs for individuals are described, ranging from specialized training for field workers and laboratory technicians, to post-graduate training leading to M.Sc., D.V.M. or Ph.D. degrees to post-doctoral positions within ILRAD's research programs. Training courses are described and details provided on application procedures. Published in English and French.

ILRAD

Research Programs at ILRAD

1985. (French edition 1986). 10 pages. 10.5 × 21.5 cm. Concertina. Free.

Research programs at ILRAD gives a brief description of ILRAD's facilities and research programs. Research is in the fields of parasitology, immunology, molecular biology, cell biology, biochemistry and veterinary pathology. Emphasis is on the development of vaccines against theileriosis and African animal trypanosomiasis, but other control methods are considered such as improving genetic resistance or the strategic use of drugs. Published in English and French.

Audiovisual Materials

S.B. Westley

ILRAD: An Introduction

1986. 150 slides. 20 minutes. Available on loan to appropriate users.

ILRAD: an introduction is a 20-minute slide-tape show produced for a general audience. It depicts the importance of livestock in Africa and the establishment and development of ILRAD. ILRAD's two target diseases are described —theileriosis and African animal trypanosomiasis — and ILRAD's research approaches are outlined. Training and information activities are mentioned, and ILRADs donors are listed.

IRRI

International Rice Research Institute

P.O. Box 933, Manila, Philippines
Cable: RICEFOUND, MANILA
Telex: 45365 RICE PM via ITT; 22456 IRI PH via RCA; 63786 RICE PN
via EASTERN

General Information

Rice is life itself to almost a third of the world's 4.5 billion people. The average annual income of rice consumers in the developing nations is less than US$300. Rice is a secondary staple for another 450 million people.

Rising populations and the demand for more food put increasing burdens on the world's small-scale rice farmers.

The International Rice Research Institute (IRRI) has as its objective the improvement of the quality and quantity of rice. Established in 1960 by the Ford and Rockefeller Foundations, this nonprofit organization is adjacent to the University of the Philippines at Los Baños, 65 km southeast of Manila.

Today, IRRI is funded through the Consultative Group for International Agricultural Research, a group of donor agencies dedicated to the improvement of agriculture in developing nations. IRRI scientists cooperate with scientists across the world to develop improved rice varieties and technology.

Since its inception, IRRI has stressed the communication of rice research and production technology through books, conference proceedings, periodicals, and other instructional materials.

Ordering instructions

Because the primary purpose of IRRI publications is to increase the flow of rice information in developing nations, *IRRI provides materials at a 70% discount in developing nations.*

Book dealers in both highly developed countries (HDC) and in less developed countries (LDC) receive a 40% discount.

Prices and airmail and surface mail costs are included for HDC and LDC. Please add appropriate postage fee to the price of each book.

CUSTOMERS IN THE PHILIPPINES MAY PAY WITH PHILIPPINE CURRENCY AT THE OFFICIAL U.S. DOLLAR-PESO EXCHANGE RATE AT THE TIME OF PURCHASE.

Surface mail orders can only be sent in 5-kg packages. We can only process surface mail orders that total 20 kg or less.

We encourage customers to order publications via *air mail* because:

1. About 30% or more of our publications dispatched via surface mail never reach their destinations, and
2. Surface mail delivery to most destinations takes 5-6 months or longer.

For bulky orders, IRRI strongly advises customers to order via *air freight.* This refers to any order of 20 kg or more (the equivalent of 8 copies of the *IRRI Annual Report,* about 475 pages each; or 32 of *Rice Improvement,* 186 pages). For air freight shipments we *only charge the customer actual air freight charges.* We are not able to process *sea freight* orders.

Prepayment is required

Payment can be made by check in Philippine pesos or US dollars *drawn on any US banks.* **IRRI cannot process checks for US dollars drawn on bank accounts outside of the USA, unless the funds are drawn through U.S. banks and this is indicated on the check.** Checks of selected currencies drawn on their respective country banks such as French francs, pound sterling, yen, HK dollars, Swiss francs, Deutchmarks, Australian dollars, Singapore dollars, are also accepted. *Checks on other currencies cannot be negotiated in the Philippines.* Checks drawn on dollar accounts in US banks by a correspondent foreign bank in special check arrangement are also acceptable. Payment may also be made in US$ cash, UNESCO coupons or traveler's checks. A list of distributing bodies of UNESCO coupons in Asia, Africa, and Latin America is found on pages 157-165.

Send orders with prepayment to:
    Communication and Publications Department
    IRRI, P.O. Box 933
    Manila, Philippines

Make checks payable to IRRI-CPD.

Basic sets

All major IRRI publications are available as *Basic Sets.*

NOTE: IRRI cannot process checks for US dollars drawn on bank accounts outside of the USA, unless the funds are drawn through U.S. banks and this is indicated on the check.

For a proforma invoice that includes exact air freight charges write to: Division F, Communication and Publications Department, IRRI, P.O. Box 933, Manila, Philippines.

Libraries or individuals in developing nations can purchase a *Basic Set* of all IRRI books and technical publications in stock (except bibliographies) for US$400.00, excluding air freight. In highly developed countries the price is US$950.00, plus air freight. A *Basic Set* is about 55 kg of publications.

Distributors

Overseas:

Customers in North America can also order IRRI books from: Agribookstore, IADS, Inc., 1611 North Kent St., Arlington, Virginia 22209, U.S.A.

Customers in **Europe** may order from: Verlag Josef Margraf, Oberwiesenstrase 32, 7000 Stuttgart 75, F. R. Germany.

Customers in **Japan** may order from: Publishers International Corporation, 2nd Newfield Building, 42-3 Ohtsuka 3-Chome Brunkku, Tokyo 112, Japan.

Customers in **Taiwan, China** may order from: Harvest Farm Magazine, 14 Wenchau Street, Taipei, Taiwan, Republic of China.

Customers in **India** may order from: Oxford Book and Stationery Company, Scindia House, New Delhi 110 001 India.
Prices are similar but postal rates are lower and delivery is faster.

Philippines:

Customers in the Philippines can also purchase IRRI publications from:

Metro Manila

- 14 Branches of National Bookstore
  Metro Manila

- IRRI Makati Office
  605 Doña Narcisa Building
  Paseo de Roxas, Makati

- Northern Foods Corporation
  3rd Floor, G&A Building
  2303 Pasong Tamo Extension
  Makati

- Philippine Education Co.
  768 Aurora Blvd.
  Cubao, Quezon City

- Heritage Book Center
  William Street
  Cubao, Quezon City

- Casalinda Bookshop
  2nd Floor San Antonio Plaza
  Forbes Park, Makati

- La Solidaridad Bookshop
  531 P. Faura
  Ermita, Metro Manila

Central Luzon

- Luz C. Nazar
  Central Luzon State University
  Muñoz, Nueva Ecija

- Pacifico T. Vizmonte, Jr.
  Department of Crop Science
  Central Luzon State University
  Muñoz, Nueva Ecija

- Gloria C. Tanguilut
  Pampanga Agricultural College
  Magalang, Pampanga

- Jacinto F. Mariano
  Wesleyan University
  College of Agriculture
  Cabanatuan City

- Robinson N. Baldomero
  Pangasinan State University
  College of Agriculture
  San Carlos City
  Pangasinan 0747

- Tarlac College of Agriculture
  Camiling, Tarlac

Ilocos Region

- N. Corpuz Enterprises, Inc.
  Paco Roman Street
  Laoag City, Ilocos Norte

- Iluminada M. Dumo
  Don Mariano Marcos Memorial State University Library
  Bacnotan, La Union

- Carlos T. Buasen
  MSAC-TEC Cooperative Inc.
  Mountain State Agricultural College
  La Trinidad, Benguet 0211

- Avelino Andrada
  Ministry of Agriculture and Food
  Vigan, Ilocos Sur

- Teodoro T. Salvador, Jr.
  Ministry of Agriculture and Food
  Region No. I, Vigan, Ilocos Sur

- Ester P. Castillo
  Bureau of Secondary Education
  Abulag National Rural & Vocational High School
  Abulag, Cagayan

- Pedro Chee, Jr.
  Ministry of Agriculture
  Municipal Agricultural Office
  San Fernando, La Union

Iloilo

- Victory D. Gabawa
  Central Philippine University
  Iloilo City

Los Bañon

- Alvin's Bookstore
  Lopez Avenue
  Los Baños, Laguna

Capiz

- Ernesto Botin
  Panay State Polytechnic College
  Mambusao, Capiz

Negros

- William Kramer
  Kramer Agro Diversification
  17th Cor. Lacson Street
  Bacolod City

Samar

- Beatriz T. Balagnir
  Librarian, Pedro Rebadulla Memorial
  Agricultural College
  Catubig, Northern Samar 7208

Visayas

- Rebecca B. Napiere
  Visca Bookstore
  Visca, Baybay, Leyte

Research Highlights

International Rice Research Institute

IRRI Highlights 1985

1986. 101 pages. 17.7 × 22.8 cm. Paperback. ISBN 971-104-131-6. HDC $13.30, LDC $4.00 plus airmail ($4.00) or surface mail ($1.00) postage.

A yearly report on IRRI progress in area such as rice breeding for resistance to diseases, insects, drought, adverse soils, and cold and hot temperature; pest control; irrigation water management; soil and crop management; cropping systems; machinery development; finances; and other items. Illustrations in color.

IRRI Highlights 1985 focuses on IRRI programs to develop improved rice varieties and management practices for farmers in economically and environmentally disadvantaged areas — those bound to rainfed upland and lowland rice, or in areas of deep water or problem soils. Simultaneously, rice scientists must continue to develop, for farmers in more favored irrigated areas, low-cost technologies to permit them to reduce production costs without sacrificing yield gains.

Also available, Research Highlights for: 1980, 1981, 1982, 1983 and 1984.

Annual Reports

International Rice Research Institute

IRRI Annual Report for 1985

1986. Approx. 500 pages. 17.7 × 25.4 cm. Paperback. ISBN 971-104-095-6. HDC $38.30, LDC $11.50 plus airmail ($9.00) or surface mail ($2.00) postage.

This is an in-depth annual report of IRRI's overall research progress presented on a problem area basis rather than by scientific discipline. Major sections include Genetic Evaluation and Utilization (problem-oriented plant breeding); Control of Diseases, Insects, and Weeds; Irrigation and Water Management; Soil and Crop Management; Environment and its Influence; Constraints on Rice Yields; Consequences of New Technology; Rice-Based Cropping Systems; and Machinery Development.

Also available, IRRI Annual Report for: 1980, 1981, 1982, 1983, and 1984.

Monographs

B.S. Vergara

A Farmer's Primer on Growing Rice

1979. 221 pages. 15 × 22.8 cm. Paperback. ISBN 971-104-051-4. HDC $6.00, LDC $2.40 plus airmail ($5.00) or surface mail ($1.00) postage.

A *Farmer's primer* was written to help progressive rice farmers and technicians understand why and how the improved rice varieties and farm technology increase production. Dr. B.S. Vergara, IRRI plant physiologist, explains agricultural practices such as why a farmer incubates seed, why he applies fertilizer, and how and when that fertilizer should be incorporated. Farmer's primer illustrations are available to encourage translations.

A farmer's primer on growing rice is now available from IRRI in:

English	— ISBN 971-104-051-4
French	— ISBN 971-104-090-5
Hiligaynon	— ISBN 971-104-091-3
Ilokano	— ISBN 971-104-092-1
Pilipino	— ISBN 971-104-093-X
Cebuano	— ISBN 971-104-118-9
Pampango	— ISBN 971-104-094-8
Warai	— ISBN 971-104-154-5
Spanish	— ISBN 971-104-125-1
Bikol	— ISBN 971-104-132-4
Kiswahili	— ISBN 971-104-135-9
Creole	— ISBN 971-104-142-1
Northern Warai	— ISBN 971-104-154-5
Maguidanao	— ISBN 971-104-155-3
Pangasinan	— ISBN 971-104-152-9

International Rice Research Institute

Field Problems of Tropical Rice: Revised Edition

1983. 172 pages. 10.2 × 17.8 cm. Paperback. ISBN 971-104-080-8. HDC $5.00, LDC $2.00 plus airmail ($3.00) or surface mail (US$1.00) postage.

Field Problems of Tropical Rice has been one of IRRI's most popular publications. The 1983 revised edition has 153 color plates to help rice workers identify common production problems such as insects, diseases, weeds, and problem soils.

Although the original edition focused on problems in Asia, the revision also identifies common problems of Africa and Latin America.

Like the original edition, the revision is designed to facilitate its low-cost translation and copublication by agricultural programs, international agencies, and publishers in developing nations. IRRI also arranges the printing of bulk orders of nonEnglish Field Problems (2,000 copies or more), using IRRI's color plates, at actual printing and paper cost, plus freight).

Field problems of tropical rice: revised edition is now available from IRRI in:

English	— ISBN 971-104-080-8
French	— ISBN 971-104-087-5
Spanish	— ISBN 971-104-086-7
Vietnamese	— ISBN 971-104-085-9
Cebuano	— ISBN 971-104-088-3
Pilipino	— ISBN 971-104-084-0
Warai	— ISBN 971-104-089-1
Pampango	— ISBN 971-104-127-8
Ilokano	— ISBN 971-104-129-4
Bengali	— No ISBN
Punjabi	— ISBN 971-104-130-7
Hiligaynon	— ISBN 971-104-143-0
Bikol	— ISBN 971-104-128-6
Nepali	— ISBN 971-104-156-1
Tamil	— ISBN 971-104-153-7
Thai	— ISBN 974-07-5425-2

IRRI is holding a limited number of nonEnglish copies for international distribution. Editions in other languages are in preparation.

International Rice Research Institute

Insights of Outstanding Farmers

1985. 114 pages. 15.24 × 22.86 cm. Paperback. ISBN 971-104- 143-X. HDC $8.30, LDC $2.50 plus airmail ($3.00) or surface mail ($1.00) postage.

"Before you teach the farmers," a Chinese peasant proverb advises, "listen to them." *Insights of outstanding farmers* contains a narration of the experiences of 14 outstanding rice farmers from 10 Asian nations who have developed packages of technology appropriate to their ecological environments and institutional settings. The farmers were honored at IRRI during the 1985 celebration of IRRI's 25th Anniversary.

One message that all of the farmers clearly conveyed is the need for additional opportunities for on-and off-farm employment and income in rural areas.

These farmers' insights are useful not only to scientists and policy makers working on rice but to all who are interested in increasing food production.

Benito S. Vergara and T. T. Chang

Flowering Response of the Rice Plant to Photoperiod

1986. 62 pages. 15.24 × 22.86 cm. Paperback. ISBN 971-104-151-0. HDC $3.30, LDC $1.00 plus airmail ($3.00) or surface mail ($1.00) postage.

The review is primarily concerned with the effect of photoperiod, or daylength, on flowering of the rice plant. The review includes discussions on how flowering response to photoperiod is affected by the phasic development of the rice plant, reception and translocation of the photoperiod stimulus, temperature, and the intensity, quality, and interruption of light. Also covered are biochemical changes during photoinduction, methods of testing photoperiod sensitivity, ecology and photoperiodism, terminologies used in describing photoperiod sensitivity, inheritance of vegetative growth duration, and problems encountered when studying photoperiod. The bibliography includes 585 papers on photoperiodism of rice; are all available through IRRI's Library and Documentation Center.

The photoperiod response of 501 specific rice cultivars tested at IRRI is included.

Dana G. Dalrymple

Development and Spread of High-Yielding Rice Varieties in Developing Countries

1986. 110 pages. 15.24 × 22.86 cm. Paperback. ISBN 971-104-159-6. HDC, send orders to METROTEC INC., 1623 Connecticut Avenue, Washington D.C. 20009, U.S.A.; LDC $3.70 plus airmail ($3.00) or surface mail ($1.00) postage.

Development and Spread of High-Yielding Rice Varieties in Developing Countries is the 7th edition of a publication formerly titled "The Development and Spread of High-Yielding Varieties of Wheat and Rice in the Less Developed Nations." This book has long been considered one of the most comprehensive and authoritative references on the adoption of semidwarf varieties. Its 6th edition was published in 1978.

The author, Dr. Dana G. Dalrymple, is an agricultural economist, U.S. Department of Agriculture, on detail to the Bureau of Science and Technology, U.S. Agency for International Development (USAID). Dr. Dalrymple serves as research advisor with the Consultative Group on International Agricultural Research (CGIAR).

USAID published the book and has provided a stock for IRRI distribution. A companion volume on wheat is available from the Bureau of Science and Technology, USAID, Washington, D.C. 20533, U.S.A.

International Rice Research Institute
(M. S. Swaminathan, Editor)

Global Aspects of Food Production

1986. 460 pages. 15.24 × 22.86 cm. Paperback. ISBN 971-104- 158-8. HDC, send orders to Tycooly Publishing, P.O. Box 213, Oxford OX2 6JS, England; LDC $4.30 plus airmail ($7.00) or surface mail ($2.00) postage.

There is abundance of global food stocks that reached nearly 300 million tons by late 1985. Thus pockets of plenty and scarcity coexist in different parts of the world. Ironically several countries that now face acute food scarcity are also those where the domestication of plants and animals took place long ago and are important centers of origin and diversity of crop plants. Why do such countries face difficulties on the food front?

Global Aspects of Food Production seeks to clarify such issues. Knowledgeable scientists discuss the multifaceted nature of the agricultural production process. They describe problems and needs as well as potentials and opportunities.

This book provides information relevant to the promotion of an international agricultural system based on principles of ecology, efficiency, and equity.

B. M. Sheppard, J. A. Litsinger, and A. T. Barrion

Helpful Insects, Spiders, and Diseases — Friends of the Farmers

1986. 132 pages. 10.2 × 17.7 cm. Paperback. ISBN 971-104- 162-6. HDC $4.30, LDC $1.30 plus airmail ($3.00) or surface mail ($1.00) postage.

This booklet illustrates representative examples of some of the more common species of predators, parasites, and diseases of insects pests of rice. It can be used with the IRRI booklet *Field Problems of Tropical Rice,* which provides information only on pest species.

Before intelligent decisions about pesticide applications can be made, it is necessary to be able to identify which insect species are pests and which are beneficial. The occurrence of beneficial organisms varies depending upon location, time of year, and crop cultural practices. Thus we made no attempt to rank the groups by their relative importance.

Scientific language is minimized to make the descriptions more easily understood. The photos provide an easy way of identifying beneficial species and thereby help prevent unnecessary chemical treatments.

Like *Field Problems,* this booklet was designed to facilitate its easy and inexpensive translation and copublication in languages other than English. IRRI does not ask for payment of royalties or payment for translations of IRRI materials published in developed nations. For details, contact the Communication and Publications Department, International Rice Research Institute, P.O. Box 933, Manila, Philippines.

International Rice Research Institute

World Rice Statistics

1986. 276 pages. 17.78 × 25.40 cm. Perfect binding. ISBN 971-104-164-2. HDC $11.30, LDC $3.40, plus airmail ($6.00) or surface mail ($2.00) postage.

This publication draws together statistics on rice production, trade, consumption, prices, and related basic information from various international and national sources. FAO publications are used as sources of data on production, area, yield, and trade of rice. Tables on rice trade include origin and destination, tariff and nontariff measures on rice, and carry-over stocks of rice. *World Rice Statistics* was prepared by IRRI Agricultural Economics Department. It is a handy reference for rice researchers and policy makers around the world.

Beth Rose

Appendix to The Rice Economy of Asia: Rice Statistics by Country, Tables with Notes

1986. 404 pages. 18 × 23.3 cm. Paperback. HDC, send orders to Resources for the Future, 1616 P. Street, N.W., Washington, D.C. 20036. U.S.A.; LDC $4.00 plus airmail ($8.00) or surface mail ($2.00) postage.

This is an appendix to the book *The Rice Economy of Asia* by Dr. R. Barker, Dr. R. W. Herdt, former IRRI economists, and Beth Rose, a scholar of Asian Affairs, Cornell University. It includes 26 tables, each containing a historical series of annual rice statistics for a specified country in Asia. The data cover the period from the turn of the century, continuing through the early 1980s. Most statistics in the tables were gathered from official government sources.

The section for each country or cluster of countries begins with a brief description of the statistical system employed by the country to gather rice production data. The tables themselves, together with notes and data sources, are presented next. The table formats are similar across countries and contain data series on total crop area; rice production, area, and yield; rice area irrigated; quantities imported and exported; the price of unmilled rice at the farm and of milled rice at the wholesale and retail levels; farm wage rates; population; and the apparent per capita availability of rice. Also included is information such as acronyms and conversion factors to assist the reader in interpreting the information.

Conferences

International Rice Research Institute

Wetland Soils: Characterization, Classification, and Utilization

1985. 559 pages. 15.24 × 22.86 cm. Paperback. ISBN 971-104- 139-1. HDC $31.30, LDC $9.40 plus airmail ($9.00) or surface mail ($2.00) postage.

Wetland Soils: characterization, classification, and utilization is the proceedings of a workshop held at the International Rice Research Institute 26 March-April 1984. The proceedings includes scientific papers and recommendations for future research and action to increase crop production from wetland soils. The workshop was sponsored by the International Network on Soil Fertility and Fertilizer Evaluation for Rice and the Soil Management Support Services of the U.S. Agency for International Development and U.S. Department of Agriculture. It brought together a unique mix of soil taxonomists and soil fertility scientists.

International Rice Research Institute

Rice Improvement in Eastern, Central, and Southern Africa

1985. 159 pages. 15.24 × 22.86 cm. Paperback. ISBN 971-104- 147-2. HDC $8.30, LDC $2.50 plus airmail ($4.00) or surface mail ($1.00) postage.

An enormous potential exists for increasing rice yields under cultivation in Africa. The present rice-growing area in eastern, central, and southern Africa is quite modest. Many governments now plan to dramatically increase rice production in response to the accelerating demand for rice by rural and urban consumers.

The International Rice Workshop held in Lusaka, Zambia 9-11 April 1984 brought together rice scientists from Kenya, Tanzania, Zimbabwe, Zambia, and Malagasy for the first time, to jointly assess their problems and formulate common strategies for rice research and development. The workshop was organized by the Department of Agriculture of Zambia and was jointly sponsored by IRRI and the International Institute for Tropical Agriculture, Nigeria.

These proceedings document the status and achievements of rice research in each national program, and directions through which future progress can be made. The workshop recommendations provide a blueprint for building a base of regional collaboration in rice research.

This volume should help alleviate the great information gap on rice in eastern and southern Africa.

International Rice Research Institute

Impact of Science on Rice

1985. 303 pages. 15.24 × 22.86 cm. Paperback. ISBN 971-104-144-8. HDC $19.70, LDC $5.90 plus airmail ($6.00) or surface mail ($2.00) postage.

Impact of Science on Rice is a compilation of papers presented during the Symposium on Rice Research: Accomplishments and Challenges, held on the occasion of IRRI's 25th anniversary. The speakers, who included ministers of agriculture and policy makers from rice-growing countries and donor nations to the Consultative Group on International Agricultural Research, discussed the many concurrent steps necessary to achieve progress in rice production. A central point of the presentations was the pivotal role of science in agricultural progress.

International Rice Research Institute

Women in Rice Farming

1985. 566 pages. 15.24 × 22.86 cm. Paperback. ISBN 566-05705-2. HDC, send orders to Gower Publishing Co. Ltd., Gower House, Croft Road, Aldershot, Hampshire, GU11 3HR, England; LDC $16.30 plus airmail ($9.00) or surface mail ($2.00) postage.

Women play a major role in rice cultivation, post-harvest processing, and marketing. Rural women work long hours in domestic and agricultural production because they often have primary responsibility for both household subsistence and child welfare. Any technology that can increase rural women's productivity, allowing them to work less and earn more, is beneficial to the welfare of rural households.

Women in Rice Farming includes papers presented at the Women in Rice Farming Systems Conference, held at IRRI in September 1983.

This volume reflects the variety of women's experiences in Asian and African rice farming. It should be of interest to agricultural economists, rural sociologists, planners concerned with international development, and those with an interest in women's issues, such as:
- Technology and the demand for women's labor in Asian rice farming;
- Wives at work-patterns of labor force participation in two rice-farming villages in the Philippines;
- The impact of new farming technology on women's employment;
- Women and technology — rice processing in Bangladesh;
- Women in African rice farming systems; and
- Two examples of emerging mechanization technologies at IRRI.

Women in Rice Farming is a joint publication of Gower Publishing Co., U.K. and IRRI. IRRI has acquired a stock of paperback copies for low-cost distribution in the Third World only.

International Rice Research Institute

Soil Physics and Rice

1985. 431 pages. 15.24 × 22.86 cm. Paperback. ISBN 971-104-146-4. HDC $16.70, LDC $5.00 plus airmail ($7.00) or surface mail ($2.00) postage.

Asian lowlands have for centuries produced enough rice to feed large populations. Food production has recently increased slightly faster than population growth — but the increases were mainly from lands with fertile soil and favorable climate. Because there is little new land to develop, rice and other crops must be grown where water is not well controlled, soils are less fertile, or there are physical constraints such as compacted soil layers.

Improved soil and water management methods are essential to increase food production from ricelands and to avoid soil erosion and degradation.

These problems were addressed by 55 scientists from 21 countries at a Workshop on Physical Aspects of Soil Management in Rice-based Cropping Systems at IRRI 10-14 Dec 1984.

Participants concluded that there is significant potential to develop practical technologies to increase food production from lowland rice-based cropping systems. Essential to these increases will be applied research on physical aspects of soil management.

This volume includes papers presented at the workshop and the participants' recommendations for research priorities.

International Rice Research Institute

Small Farm Equipment for Developing Countries

1986. 637 pages. 15.24 × 22.86 cm. Paperback. ISBN 971-104- 157-X. HDC $30.00, LDC $9.00 plus airmail ($10.00) or surface mail ($3.00) postage.

To take advantage of improved rice varieties and farm practices, small farmers need to mechanize certain operations of crop production.

But constraints such as small landholdings, excess labor in the agricultural sector, and a lack of capital limit the adoption of mechanization in developing countries. Above all, mechanization should not create serious problems where rural labor is plentiful. The impact of agricultural mechanization on female labor needs particular attention. The aim should be to reduce drudgery, improve effiency and productivity, and generate diversified opportunities for labor employment.

Eighty-seven research workers from 22 countries gathered at IRRI 2-7 September 1985 for the conference "Small Farm Equipment for Developing Countries: Past Experiences and Future Priorities."

Small Farm Equipment for Developing Countries includes the papers and recommendations of that conference.

International Rice Research Institute

Rice Grain Quality and Marketing

1986. 76 pages. 21.59 × 27.94 cm. Paperback. ISBN 971-104-149-9. HDC $5.00, LDC $1.50 plus airmail ($4.00) or surface mail ($1.00) postage.

Grain quality becomes increasingly important as more Asian countries become self-sufficient in rice and look toward export markets for selling surpluses. Although production, harvesting, and postharvest operations affect quality of milled rice, the variety remains the most important determinant of market quality. Consumers favor certain varieties and value specific appearances and tastes of milled rice for home cooking.

Recognizing that markets are most responsive to high quality rice, planners of the 1985 International Rice Research Conference, 1-5 June at IRRI included a half-day session on grain quality and marketing. Presentations covered the status of rice quality in world and domestic markets, consumer demand for rices with certain physical and chemical characteristics, effects of environment and variety on milling quality, and breeding for rices of excellent processing, cooking, and eating qualities.

Participants also discussed priority problem areas in rice breeding for market quality. Recommendations stress research needs in milling quality and consumer preferences.

Rice Grain Quality and Marketing includes the papers and recommendations of that meeting.

International Rice Research Institute

Progress in Upland Rice

1986. 578 pages. 15.24 × 22.86 cm. Paperback. ISBN 971-104- 150-2. HDC $25.00, LDC $7.50 plus airmail ($9.00) or surface mail ($2.00) postage.

The second International Upland Rice Conference was held 4-8 March 1985 in Jakarta, Indonesia. The conference was divided into four technical sessions: 1) characterization and classification of upland rice-growing environment, 2) integrated upland rice farming systems, 3) biological stresses with emphasis on blast disease, and 4) ecological, social, economic, and technical issues of preproduction testing and production.

Progress in Upland Rice is the proceedings of that conference. Recommendations for research priorities and strategies, as well as conference papers, are included.

The Jakarta Conference was organized by the Upland Rice Intercenter Research Coordinating Committee and the Agency for Agricultural Research and Development of Indonesia.

International Rice Research Institute

Rice Genetics: Proceedings of the International Rice Genetics and Cytogenetics Symposium

1986. 956 pages. 15.24 × 22.86 cm. Paperback. ISBN 971-104- 148-0. HDC $23.30, LDC $7.00 plus airmail ($20.00) or surface mail ($5.00) postage.

A continuous updating of knowledge of the genetics of the rice plant and the precise identification of the nature of the genetic control of important economic characters have become particularly urgent in the context of recent advances in molecular biology and genetic engineering.

Two hundred one scientists from 31 countries met at IRRI in June 1985 for a symposium on "Rice Genetics and Cytogenetics" — the only such symposium held since 1963. The state of knowledge of rice genetics was reviewed, including recent advances in fields such as gene mapping, identification and conservation of genetic stocks, and gene symbolization and nomenclature.

This volume includes papers presented at the symposium.

International Rice Research Institute

Physical Measurement in Flooded Rice Soils: The Japanese Methodologies

1986. 65 pages. 15.24 × 22.8 cm. Paperback. ISBN 971-104- 163-4. HDC $3.30, LDC $1.00 plus airmail ($4.00) or surface mail ($1.00) postage.

Worldwide demand for rice will increase by 3% annually for the next 15 years. The increased production must be met by growing rice on lands where water is not well controlled, or soils are less fertile, or there are physical constraints such as compacted soil layers. Improved methods of soil and water management for ricelands must be developed and adopted, both to increase food production and to avoid soil erosion and land degradation.

Participants at a 1985 workshop on "Physical Aspects of Soil Management in Rice-based Cropping Systems" recommended strongly that the existing Japanese methodologies of soil-physical measurement in flooded rice soils, developed over years of careful research, be published in English. Ten authors from Japanese universities and the Japanese National Research Institute of Agricultural Engineering wrote and translated the 10 component chapters of this book that describe measuring methods relating to soil texture, structure, mechanics, and water.

Miscellaneous

Technical Advisory Committee

Training in the CGIAR System

1986. 136 pages. 18.41 × 27.76 cm. Paperback. ISBN 971-104- 161-8. HDC $10.00, LDC $3.00 plus airmail ($4.00) or surface mail ($1.00) postage.

The International Agricultural Research Centers emphasize training as an essential thrust to generate, promote, and disseminate research results. This publication describes efforts to forge and strengthen the partnership of the Consultative Group on International Agricultural Research and the nations it serves in its training endeavors to translate knowledge into more and better food from farmers' fields. It is based mainly on the report of a study team commissioned by the Technical Advisory Committee (TAC) to the CGIAR to review progress in training at the Centers and to suggest ways to further enhance this progress.

IRRI arranged the printing of the publication and assists the CGIAR and TAC in its distribution.

Bibliographies

Requests and purchases for bibliographies may be addressed to:
        Library and Documentation Center
        IRRI, P.O. Box 933
        Manila, Philippines

Milagros C. Zamora, The Library and Documentation Center

International Bibliography of Rice Research (Annual Supplements)

1985. 736 pages. 21.7 × 29.9 cm. Paperback. ISSN 0074-2021. HDC $93.00, LDC $28.00 plus airmail ($28.00) or surface mail ($2.25) postage.

This annual bibliography contains worldwide references to rice research published each year. The arrangement is classified. An author index and a manually produced keyword index are provided.

Also available, annual supplement for: 1977, 1978, 1979, 1980, 1981, 1982, 1983, and 1984.

Periodicals

International Rice Research Institute
IRRI Research Paper Series (IRPS)

Number of pages varies. 21.5 × 27.9 cm. Saddle stitched. ISSN 0115-3862.

The *IRRI Research Paper Series* is a vehicle for timely publication of research findings by IRRI and associated scientists. The papers go regularly to libraries and certain institutions. Individuals may request *for only one or two selected papers* on a complimentary basis. Requests for complimentary copies must be made on organizational letterheads. Additional copies may be purchased at prices indicated below:

Overseas				Philippines	
Per copy		Air-mail	Surface mail	Air-mail	Surface mail
HDC	LDC				
$2.00	$0.60	$3.00	$1.00	₱5.40	₱1.00

Bound volumes of IRPS, 20 issues/volume (IRPS 1-20, IRPS 21-40, IRPS 41-60, IRPS 61-80, IRPS 81-100) are available at the following prices:

Overseas				Philippines	
Per copy		Air-mail	Surface mail	Air-mail	Surface mail
HDC	LDC				
$73.50	$23.50	$13.00	$2.10	₱38.00	₱7.00

No. 111 Changes in rice breeding in 10 Asian countries: 1965-84
No. 112 Design parameters affecting the performance of the IRRI-designed axial-flow pump
No. 113 Boron toxicity in rice
No. 114 Energy analysis, rice production systems, and rice research
No. 115 Production risk and optimal fertilizer rates: an application of the random coefficient model
No. 116 Consumer demand for rice grain quality in Thailand, Indonesia, and the Philippines
No. 117 Morphological changes in rice panicle development

Also available: IRPS No. 1-110.

International Rice Research Institute

International Rice Research Newsletter (IRRN)

Number of pages varies. 21.5 × 27.9 cm. Saddle stitched. ISSN 0115-0944. Free.

The *International Rice Research Newsletter* (IRRN) is published by IRRI in the interest of improving the communication of relevant rice research among scientists in many nations. The IRRN is published every 2 months and has an annual subject index.

The IRRN invites all scientists to contribute concise summaries of significant research on rice or rice-based cropping systems for review. Contributions should be limited to no more than two double-spaced pages and no more than two graphics (tables, figures, or photographs). Authors are identified by name, title, and research organization.

International Rice Research Institute

IRRI Reporter

Usually 4 pages. 21.5 × 27.9 cm. Single fold. ISSN 0115-2467. Free.

The *IRRI Reporter* is a quarterly publication describing IRRI's research program and directions.

Complimentary subscriptions to the *IRRI Reporter* and the IRRN are available.

ISNAR

International Service for National Agricultural Research

P.O. Box 93375, 2509 AJ - The Hague, Netherlands
Cable:ISNAR
Telex: 33746
Telephone: 070/47.29.91

Mandate and Objectives of ISNAR

ISNAR's primary mandate is "to help strengthen national agricultural research capabilities in developing countires." Only if the scientific capacity and research management capability of the countries concerned are adequately developed so as to enable them to test, adapt, and complement the newly developed technologies, can the ultimate aim of the CGIAR system — of helping the developing countries achieve self-sustaining technical and economic growth — be reached.

ISNAR concentrates its assistance to developing countries mainly on program, priorities, policy, organizational and management programs needed to improve the performance of national agricultural research systems. It is concerned with commodities and other renewable natural resources important to national development objectives including, but not limited to, the food commodities covered by other components of the CGIAR system.

Such assistance is available to any developing country upon request, ISNAR's available staffing and financial limitations permitting. The goal is to enable developing countries to plan, organize, manage, and execute agricultural researh more effectively from their own human, natural, and financial resources as soon as possible. The service seeks to complement, and not compete with, other sources of technical assistance and frames its program to encourage full use of technical expertise from other sources acceptable to the government concerned.

Annual Reports

Annual Report 1985.

June 1986. 18.4 × 24 cm. Sewn. Single copy free. Multiple copies US$5.00.

Detailed full reports on ISNAR activities during one year.

ISNAR Documents

Serving National Agricultural Research Systems: Lessons from Country Experiences. 1980-84.

Conferences

Les Besoins de Formation Continue en Gestion et en Méthodologies de la Recherche a l'Institut des Sciences Agronomiques du Rwanda. Juin 1985.

Women and Agricultural Technology: Relevance for Research. Volume 1 - Analyses and Conclusions. July 1985.

Women and Agricultural Technology: Relevance for Research. Volume 2 - Experiences in International and National Research. July 1985.

Agricultural Research Management - Nkolbisson, 1984. August 1985.

Le Management de la Recherche Agricole - Nkolbisson, 1984. Août 1985.

Research Management

Considérations pour le Développement des Capacités de Recherche Agricole Nationale en Soutien du Développement Agricole. Mai 1986. Traduction de l'original anglais.

Reprints

ISNAR Reprint Series No.3. International Technology Transfer. May 1986.

Proagro

PROAGRO Paper No. 3. Las Funciones del Sector Público en el Mejoramiento Genético de los Principales Cultivos de la Región Pampeana en Argentina. Julio 1985.

PROAGRO Paper No. 4. Articulaciones Sociales y Cambios Técnicos en el Agro Ecuatoriano. Julio 1985.

PROAGRO Paper No. 5. Los Hitos Tecnológicos en la Agricultura Pampeana. Agosto 1985.

PROAGRO Paper No. 6. A Relacao Setor Público Privado Na Geraçao de Tecnologia Agrícola No Brasil. Janeiro 1986.

PROAGRO Paper No. 7. La Fundación Servicio para el Agricultor en el Sistema Agríçola Venezolano. Enero 1986.

PROAGRO Paper No. 8 La Industria de Semillas en Países Semi-Industrializados: Los Casos de Argentina y Brasil. Enero 1986.

PROAGRO Paper No. 9. Política Tecnológica Agropecuaria y Desarrollo del Sector Privado: El Caso de la Región Pampeana Argentina. Enero 1986.

PROAGRO Paper No. 10. The Development of the Private Sector in Agricultural Research: Implications for Public Research Institutions. January 1986.

Working Papers

Working Paper No. 3. Analyzing Conditions of Service for Agricultural Researchers: An Experiment Using Earnings Functions. July 1985.

Working Paper No. 4. Agricultural Research Organization in the Developing World: Diversity and Evolution. March 1986.

Country Reports

Africa

Kenya Agricultural Research Strategy and Plan December 1985. Volume 1 - Organization and Structure. Volume 2 -Priorities and Programs.

Requirements for Strengthening Postgraduate Research Training in Agriculture and Veterinary Medicine: A Contribution to Kenya's Manpower.
Development for Agricultural Research. July 1985.

Asia and South Pacific

Agricultural Research Plan - Fiji. June 1985.

Latin America and Caribbean

Working to Develop Support at the Political Level for National Agricultural Research: the Case of the Dominican Republic. June 1985.

Regional Research Networks: the Experience of PRECODEPA.
CIP, ISNAR. November 1985.

WARDA

West Africa Rice Development Association

P.O. Box 1010, Monrovia, Liberia
Cable: WARDA; Telex: WARDA 4333 LI
Telephone: 22.14.66 & 22.19.63

General Information

The West Africa Rice Development Association which was formed in September 1970 under the auspices of UNDP, FAO and ECA is an intergovernmental organization consisting of 16 member countries namely: Benin, Burkina-Faso (Upper-Volta), Chad, Ivory Coast, The Gambia, Ghana, Guinea, Guinea-Bissau, Liberia, Mali, Mauritania, Niger, Nigeria, Senegal, Sierra Leone and Togo.

The Headquarters of WARDA are located in Monrovia (Liberia) while its four Regional Research Stations are located in Bouake, Ivory Coast (Upland rice); Rokupr, Sierra Leone (mangrove and swamp rice); Richard-Toll, Senegal (irrigated rice); and Mopti, Mali (deep water/ floating rice). In addition, WARDA has a Regional Training Center at Fendall (Liberia).

WARDA is one of the 13 International Agricultural Research Centers (IARCs) of the Consultative Group on International Agricultural Research (CGIAR). The Association has been mandated to assist its member countries to achieve self-sufficiency in rice, a staple food of West Africans.

AVRDC

The Asian Vegetable Research and Development Center

P.O. Box 42, Shanhua, Tainan, 74199
Taiwan, Republic of China
Cable: ASVEG Shanhua
Telex: 73580 AVRDC
Telephone: (06)5837801

About AVRDC

AVRDC is the international research and training center charged with improving vegetable crop production in the tropics. Its multinational scientific staff conducts research aimed at increasing the production potential of tomato, Chinese cabbage and sweet potato as well as soybean and mungbean. The Center's research agenda also includes the study of the nutritional, environmental and management factors that influence vegetable production in the tropics. Emphasis is similarly directed to scientific and training programs conducted in collaboration with national and international research organizations, agricultural development agencies and institutions of higher education.

How to Order

The publications listed below can be ordered from the AVRDC Office of Information Services (OIS). Most are available free of charge and will be posted upon request via surface mail. If airmail delivery or multiple copies are required, please check the price list located below. All charges should be paid with a US dollar check payable by a bank in the US, made out to AVRDC.

Notice to Librarians

Single copies of all AVRDC publications are available to libraries free of charge. If more than one copy is required, or if our records show that your library received a free copy in the past, you will be charged for the publication at the "additional copies" price listed below. All library copies will be sent by surface mail unless otherwise indicated in your request. Please consult the price list for airmail charges.

Research Highlights

Bruce T. McLean (Compiler/Editor)
AVRDC Progress Report Summaries 1985

Sept. 1986. _____ pages. 18 × 25.5 cm. Soft cover. ISSN 0258- 3097. Free (includes surface mail postage). Airmail postage, $3.00.

Provides a brief summary of all major activities conducted by AVRDC in 1985. It also contains considerable institutional information about AVRDC.

Bruce T. McLean (Compiler/Editor)

1985 AVRDC Highlights

1986. 8 pages/folds. 9.5 × 21.2 cm. Folded. Free (includes surface mail postage). Airmail postage, $0.50.

This short color brochure contains the highlights (illustrated) of AVRDC's research activities in 1985.

Available in: English, Chinese, French, German, Indonesian, Italian, Japanese and Korean.

Annual Reports

AVRDC Progress Report 1985

Dec. 1986. 450 pages. 19 × 26 cm. Soft cover. ISSN 0258- 3089. HDC $10.00, LDC $10.00 plus airmail ($10.00) or surface mail ($3.50) postage.

Provides a detailed report of all major activities conducted by AVRDC in 1985.

Conference, Symposia, and Workshop Proceedings

N. S. Talekar (Scienfitic Editor), T. D. Griggs (Publications Editor).

Diamondback Moth Proceedings

1986. 483 pages. 17.7 × 25.3 cm. Soft cover. ISBN 92-9058- 0021-8. HDC $18.00, LDC $18.00 plus airmail ($10.00) or surface mail ($3.50) postage.

Contains 40 reports on all aspects of this serious cruciferous pest from the International Workshop on Diamondback Moth Management, 11- 15 March 1985.

S. Shanmugasundaram (Scientific Editor), P. Lastimosa (Publications Editor).

Soybean Varietal Improvement Workshop

Dec. 1986. _____pages. 17.5 × 25.3 cm. Soft cover. ISBN 92- 90580- 0023-5. HDC $7.50, LDC $7.50 plus airmail ($2.50) or surface mail ($1.00) postage.

Contains 10 reports given during the 1984 Soybean Varietal Improvement Workshop.

Miscellaneous

P. Van der Goot
Agromyzid Flies of Some Native Legume Crops

Sept. 1984. 104 pages. 17 × 25 cm. Soft cover. ISBN 92-9058-0006-4. HDC $5.00, LDC $5.00 plus airmail ($2.50) or surface mail ($1.00) postage.

This publication was translated from the Dutch "De Agromyza-Vliegjes der Inlandsche Katjang-Gewassen op Java" of 1930.

S. Shanmugasundaram, T. S. Toung and L. F. Chen
AVRDC Soybean Evaluation Trial (ASET)

Dec. 1985. 66 pages. 21 × 27.51 cm. Soft cover. Free. Airmail postage, $2.50, surface mail postage, $1.00.

First report of results from international network of AVRDC Soybean Evaluation Trial (ASET) cooperators from 1980-81.

Chang-Soon Ahn, Jin-Hu Chen, Hao-Koan Chen
International Mungbean Nursery

Dec. 1985. 94 pages. 21 × 27 cm. Soft cover. Free. Airmail postage, $2.50, surface mail postage, $1.00.

Performance of the ninth (1981) and tenth (1983) International Mungbean Nursery (IMN) trials conducted by AVRDC's international cooperators.

George Kuo
Handling of Sweet Potato Germplasm

June 1985. 2 pages. 20.4 × 26 cm. Free.

This International Cooperator's Guide lists the two forms which AVRDC distributes sweet potato germplasm without spreading insects or diseases.

George Kuo, Steve Lin, and Sylvia Green
Sweet Potato Germplasm for International Cooperators

June 1985. 4 pages. 20.4 × 26 cm. Free.

This International Cooperator's Guide lists the 30 AVRDC cultivars currently available and provides information on how they can be obtained.

John S. Caldwell

Assessing Rainy Season Vegetable Production Alternatives: A Case Study in "Upstream" Farming Systems Research

Oct. 1986. 50 pages. 17.5 × 25 cm. Soft cover. ISBN 92-9058- 0024-3. Free (includes surface mail postage). Airmail postage, $1.00.

The study was conducted with the goal of developing a methodology for integrating surveys of vegetable production, consumption and marketing with experiments based on AVRDC crop management research.

ICIPE

The International Centre of Insect Physiology and Ecology

P.O. Box 30772, Nairobi, Kenya
Cable: ICIPE, Nairobi
Telex: 22053
Telephone: Nairobi 43049/43081/43719

Introduction

The International Centre of Insect Physiology and Ecology (ICIPE) was established in 1970 and became functional in 1972. The idea was inspired by the realization that long-term solutions to pest problems in the tropics could not rely on chemical insecticides alone, and that even the effective use of these depended on a thorough understanding of insect biology, ecology and related disciplines.

Research Programmes

ICIPE's Research Programme is primarily concerned with tropical insect pests of crops and vectors of livestock diseases. The crop pests under investigation include the crop borers of maize, sorghum, and cowpeas. The livestock pests being studied are tsetse, the vector of trypahosomiasis, and ticks, the vector of several debilitating cattle diseases including East Coast Fever. In earlier years, there was also substantial emphasis on insect vectors of human diseases. All that remains of this medical vectors research is a single project on leishmaniases, which is concerned with the sandfly vector of the disease, which is important in the semi-arid rural areas. The ultimate impact of ICIPE's research will be the development of integrated pest and vector management systems for these economically important pests and disease vectors.

ICIPE's research programmes are supported by four research units which provide assistance in chemistry, fine structure, sensory physiology and biostatistics.

Training

Training at the ICIPE is conducted in three major areas:
(a) Professional scientific training is accomplished at both the postdoctoral and postgraduate levels. The latter is done in close collaboration with several African universities. (b) Short-term training courses in pest and vector management are offered to practising insect scientists and technologists from the developing world. (c) There is programme for the upgrading of special skills and techniques of the staff of the ICIPE itself.

ICLARM

International Center for Living Aquatic Resources Management

MCC P.O. Box 1501, Makati, Metro Manila
Cable Address: ICLARM Manila
Telex Address: ITT 45658 ICLARM PM
Telephone: 818-04-66; 818-92-83

Introduction

ICLARM was organized in 1975 in response to a broadly perceived need to strengthen research on tropical aquatic resources for the benefit of developing countries. It was recognized that development of aquaculture and of viable fisheries management practices is often hindered by the inadequacy of technical information pertinent to problems encountered in the tropics. ICLARM's research has been oriented toward the goal of improving the condition of the rural poor in developing countries by improving their incomes, employment opportunities and productivity through the wise use of aquatic resources. ICLARM was incorporated in the Philippines in 1977. Since then ICLARM has played an important role in fisheries development and has been widely recognized for its research contributions.

Objectives

The objectives of ICLARM as stated in its Articles of Incorporation may be summarized as follows:
- To conduct and assist with research on fish production management, presentation, distribution, and utilization to assist peoples of the world in meeting their nutritional and economic needs.
- To improve the efficiency of culture and capture fisheries through coordinated research, education and training, linked with appropriate development and extension programs.
- To upgrade the social, economic, and nutritional status of people in less-developled areas through improvement of small-scale fisheries.
- To encourage labor intensive and low-energy input systems where appropriate.
- To publish and disseminate research findings in support of the Center's objectives.
- To organize and conduct conferences, forums, and workshops for discussion of current problems and for exchange of research results.

Annual Reports

J. L. Maclean, L. B. Dizon, and M. S. M. Sadorra

ICLARM Report 1985

1986. 117 pages. 18 × 25.5 cm. Perfect binding. ISSN 0115-4494, ISBN 972-1022-23-0. Free plus air mail $6.00 postage.

Conferences

I. R. Smith, E. B. Torres and E. O. Tan

Philippine Tilapia Economics

1985. 261 pages. 18 × 25.5 cm. Perfect binding. ISSN 0115-4435, ISBN 971-1022-18-4. Air mail $26.25; surface mail $13.75.

Contains 18 papers presented at the PCARRD-ICLARM Workshop on Tilapia Economics. 10-13 August 1983, at Laguna, Philippines together with working group reports and recommendations of the workshop.

S. Tookwinas (translated by E. W. McCoy)

Commercial Cockle Farming in Southern Thailand

1985. 13 pages. 21.5 × 28 cm. Stapled. ISSN 0115-4141, ISBN 971-1022-20-6. Air mail $2.50; surface mail $1.25.

Describes commercial cockle culture in Satun Province, Southern Thailand. Original in Thai published in 1983.

C. L. Angell

The Biology and Culture of Tropical Osyters

1986. 42 pages. 21.5 × 28 cm. Saddle stitched. Air mail $6.00; surface mail $3.00.

Reviews the biology, ecology and culture techniques both experimental and commercial, used in the tropics; describes problems associated with tropical oyster farming; and points out research needs to develop this form of aquaculture further. Oyster genera discussed are: Ostrea, Crassostrea and Saccostrea. Discusses also the advantages and disadvantages of various species of each genus with regard to aquaculture.

M. J. Broom

The Biology and Culture of Marine Bivalve Molluscs of the Genus *Anadara*

1985. 37 pages. 21.5 × 28 cm. Saddle stitched. ISSN 0115-4389, ISBN 971-1022-21-4. Air mail $7.50; surface mail $3.75.

A review of the general biology, ecology, population dynamics, reproduction and culture methods of marine bivalves of the family Arcidae, sub-family Anadarinae. These cockles are harvested on a subsistence basis in many tropical, subtropical and warm temperate areas. Important species are *Anadara granosa* (L.), *A. subcrenata* (Lischke) and *A. broughtoni* (Schrenk).

Bibliographies

J. L. Munro and W. J. Nash

A Bibliography of the Giant Clams

1985. 26 pages. 21.5 × 28 cm. Saddle stitched. ISSN 0115-5997, ISBN 971-1022-22-2. Air mail $5.65; surface mail, $3.15.

A bibliography on Tridacnidae (around 300 entries) of scientific papers and reports that deal with aspects of biology, ecology, exploitation and cultivation of the living species of giant clams.

Periodicals

NAGA, The ICLARM Quarterly

January, April, July, October 1986. No. 1, 24 p.; No. 2, 32 p.; No. 3, 24 p. 21.5 × 28 cm. Saddle stitched. ISSN 0116-290X. HDC $14.00 (air mail postage included); LDC free (surface mail).

New name for *ICLARM Newsletter* starting January 1986. Volume numbering is continuous from the ICLARM Newsletter.

ICRAF

International Council for Research in Agroforestry

P.O. Box 30677, Nairobi, Kenya
Cable: ICRAF-
Telex: 22048
Telephone: 29867

General Information

ICRAF is an autonomous, nonprofit international council governed by a board of trustees with equal representation from developed and developing countries. With the exception of a representative of the host country, Kenya, trustees do not represent countries or organization but are elected on individual merit. ICRAF derives its financial support from governments and international, private, and public organizations and agencies.

ICRAF's objective is to improve the nutritional, economic, and social well-being of the peoples of developing countries by promoting agroforestry systems for better land use without harming the environment. ICRAF acts as an international catalyst in agroforestry research.

ICRAF's program of work includes
- the development of methodologies to identify social, economic, and ecological constraints in land-use systems and to assess the potential of agroforestry technologies to overcome such constraints;
- the systematic collation and assessment of agroforestry knowledge and the development of methods of studying and evaluating agroforestry technologies; and
- the efficient dissemination of methodologies and knowledge to scientists and development planners in the tropical and sub-tropical developing world.

In this catalog ICRAF publications are listed in the following categories:

Books, proceedings, and reviews
Science and practice of agroforestry
ICRAF communications
ICRAF reprints
ICRAF working papers
Agroforestry systems
About ICRAF
Miscellaneous papers

Annual Report

ICRAF

Annual Report 1985

1986. 44 pages. Nairobi. (Limited Circulation).

ICRAF Reprints

The ICRAF reprints series, issued under the council's own imprint, consists of articles by ICRAF Staff or papers by other scientists first published elsewhere but as a result of the council's own research activities.

IR-22 1985 Lundgren, Bjorn. **Global Deforestration, Its causes and suggested remedies.** Reprinted from 'Agroforestry Systems' 3: 91-95.

IR-23 1985 Nair, P.K.R. **Classification of Agroforestry Systems.** Reprinted from 'Agroforestry Systems' 3: 97-128.

IR-24 1985 Huxley, P.A. **Experimental Agroforestry - Progress through Perception and Collaboration?** Reprinted from 'Agro-Forestry Systems' 3: 129-138.

IR-25 1985 Von Carlowitz, Petor G. **Some Considerations Regarding Principles and Practice of Information Collection on Multipurpose Trees.** Reprinted from 'Agroforestry Systems' 3: 181-195

IR-26 1985 Huxley, Peter A. **Systematic Designs for Field Experimentation with Multipurpose Trees.** Reprinted from 'Agroforestry Systems' 3: 197-207.

IR-27 1985 Lundgren, B. and P.K.R. Nair. **Agroforestry for Soil Conservation.** Reprinted from 'Soil Erosion and Conservation'. pp. 703-717.

84-28 1985 Baumer, Michel. **Amenagement de la Nature Dans Les Regions Arides.** Reprinted from 'Amenagement et Nature' No. 78.

IR-29 1986 Huxley, Peter A. **Tree/Crop Interface - or Simplifying the Biological/Environmental Study of Mixed Cropping Agroforestry Systems.** Reprinted from 'Agroforestry Systems' 3: 251-256.

IR-30 1986 Young, A. **Land Evaluation and Agroforestry Diagnosis and Design: Towards a Reconciliation of Procedures.** Reprinted from 'Soil Survey and Land Evaluation' 5: 61-76.

IR-31 1986 Young, A. **Effects of Trees on Soils.** Reprinted from 'Amelioration of Soils by Trees'. London: Commonwealth Science Council, 1986. pp. 28-41.

IR-32 1986 Raintree, J.B. and KM. Warner. **Agroforestry Pathways for the Intensification of Shifting Cultivation.** Reprinted from 'Agroforestry Systems' 4: 39-54.

ICRAF Working Papers

Working papers are made available in limited numbers for comment and discussion and to inform interested colleagues about work in progress at ICRAF. Comments and suggestions are invited, and they should be directed to the author (s). Material in working papers may be cited but working papers may not be reproduced without permission.

IWP-33 1985 Nair, P.K.R. **Agroforestry in the Context of Land Clearing and Development in the Tropics.** Nairobi: ICRAF. 41 pp. + Appendix.

IWP-34 1985 Young, Anthony. **The Potential of Agroforestry as a Practical Means of Sustaining Soil Fertility.** Nairobi: ICRAF. 28 pp.

IWP-35 1985 Fernandes, Erick C.M. **Considerations for the Planning, Implementation, and Evaluation of On-Farm Experimentation in Agroforestry Farming Systems.** Nairobi: ICRAF. 18 pp.

IWP-36 1985 Rochelau, Dianne E. **Land Use Planning with Rural Farm Households and Communities: Participatory Agroforestry Research.** Nairobi: ICRAF. 43 pp.

IWP-37 1985 Rocheleau, Dianne E. **Criteria for Re-Appraisal and Re-Design: Intra-Household and Between-Household Aspects of FSRE in three Kenyan Agroforestry Projects.** Nairobi: ICRAF. 21 pp. $\times$ Annexture.

IWP-38 1986 Fernandes, E.C.M. and P.K.R. Nair. **An Evaluation of the Structure and Function of Tropical Homegardens.** Nairobi: ICRAF. 32 pp. + Annexture.

IWP-39 1986 Raintree, J.B. and F. Torres. **Agroforestry Research in Farming Systems Perspective: The ICRAF Approach.** Nairobi: ICRAF. 30 pp.

IWP-40 1986 Huxley, P.A. **Rationalising Research on Hedgerow Intercropping - An Overview with some Comments and Suggestions.** Nairobi: ICRAF. 71 pp.

IWP-41 1986 Von Carlowitz, P.G. **Recommendations for the Design and Establishment of Demonstration Trials at the Ethiopian Centre for Community Forestry and Soil Conservation.** Nairobi: ICRAF. 89 pp.

Newsletters

NL 1986. No. 17. 8 pp. May (EN, FR, ES).
No. 18. 8 pp. September (EN, FR, ES).

WINROCK

WINROCK International Institute for Agricultural Development

Route 3, Petit Jean Mountain, Arkansas 72110, U.S.A.
Telex: 910-720-6616 WI HQ UD
Telephone: (501)-727-5435
Dialcom: 41:TCN400

Introduction

The Winrock International Institute for Agricultural Development was established in 1985, and incorporates more than 50 years of combined organizational experience in providing technical assistance and professional expertise to international agricultural development. One of the largest independent entities engaged in international agricultural development, Winrock International was created through the merging of the Agricultural Development Center (A/D/C), the International Agricultural Development Service (IADS), and the Winrock International Livestock Research and Training Center. Winrock International is an autonomous, nonprofit corporation.

Mission

Winrock International's mission is to help people around the world by increasing agricultural productivity and improving nutrition. Since agricultural development is by its nature a team effort, cooperative relationships are the key to its success. Accordingly, Winrock staff members work with host governments, private voluntary organizations, producer groups, and others to provide research and analysis, education and training, technical assistance, and communications services. Areas of emphasis are human resources, renewable resources, food policy, animal agriculture and farming systems, and agricultural research and extension.

Monographs

Agricultural Development Council

Food Policy Analysis in Thailand

1985. 347 pages. 18 × 25 cm. Perfect binding. HDC $16.00, LDC $8.00 plus airmail ($16.30) or surface mail ($2.10) postage.

The six main studies and five shorter essays in this volume resulted from the Food Policy Analysis Project in Thailand. Thailand is an important food exporter, yet it experiences 1) poverty and malnutrition despite impressive growth rates in agricultural output and rising food exports; 2) stagnant crop yields despite the availability of high-yielding varieties, impressive irrigation investments, and the farmers' reputed responsiveness to economic incentives; and 3) policies affecting the performance of the food sector formulated with no policy analysis input despite the increasing numbers of economists in the country with sophisticated analytical tools.

Winrock International Institute for Agricultural Development
Agroecosystem Analysis for Research and Development

1986. 111 pages. 18.5 × 26.5 cm. Perfect binding. HDC $6.00, LDC $3.00 plus airmail ($6.25) or surface mail ($1.00) postage.

The Green Revolution, with its narrow emphasis on productivity, largely ignored both environmental and socioeconomic heterogeneity. The author of this publication presents an innovative approach, agroecosystem research and development, to the complexities of agricultural development. This approach can be used within the framework of farming systems research or integrated rural development, and is based on the disciplines of agricultural and human ecology.

Conference, Workshop, and Symposia Proceedings

Winrock International Institute for Agricultural Development
Research and Development of Vegetables in the Tropics

1986. 88 pages. 21 × 28 cm. Perfect binding. ISBN 0-933595-04-2. HDC $11.00, LDC $5.50 plus airmail ($6.30) or surface mail ($1.00) postage.

This collection of presentation summaries resulted from a meeting of 25 agricultural specialists who explored ways to help the developing countries of the tropics to improve the diets and incomes of their poor by promoting vegetables. The group recommended the development of network of organizations and individuals concerned with research and development of vegetables in the tropics and the formation of a small unit to provide scientific and administrative leadership and coordination.

Winrock International Institute for Agricultural Development
Multispecies Grazing

1985. 235 pages. 21 × 28 cm. Perfect binding. ISBN 0-933595-02-6. HDC $10.00, LDC $5.00 plus airmail ($11.65) or surface mail ($1.75) postage.

This proceedings document contains the deliberations of 30 conferees from a variety of disciplines. The conference was held in response to inquiries as to whether combining sheep and cattle would improve the economic efficiency of miswestern U.S. farms. Research data and producer experiences for both arid and humid areas of the country were reviewed, evaluated, and summarized.

Agricultural Development Council

Irrigation Management: Research from Southeast Asia

1981. 248 pages. 15 × 23 cm. Perfect binding. HDC $12.50, LDC $6.25 plus airmail ($6.25) or surface mail ($1.00) postage.

Most of the papers in this volume pay much-needed attention to the farmers because it is believed that greater farmer involvement in the design and management of irrigation systems will lead to better performance in these systems. The primary objective of this book, written mainly by Asians for Asians, is to place a set of discrete studies within easy reach of research scholars, students, and practitioners — those who must make policy choices regarding the design and management of irrigation systems.

Agricultural Development Council

Farm Power and Employment in Asia

1984. 437 pages. 15 × 21 cm. Perfect binding. HDC $17.00, LDC $8.50 plus airmail ($13.20) or surface mail ($1.80) postage.

This collection discusses the history, impact, and future of the mechanization of small farms in Sri Lanka, Bangladesh, Indonesia, and Thailand. The findings of the four studies show how various types of mechanization offer private and social advantage in some situations but not in others. Editorial introductions and conclusions to each section help to mold the papers into a cohesive whole.

Miscellaneous (Handbooks)

Winrock International Livestock Research and Training Center

Goat Health Handbook

1983. 123 pages. 14 × 20 cm. Comb. HDC $4.25, LDC $4.25 plus airmail ($3.95) or surface mail ($1.00) postage.

This field handbook was designed to assist agricultural development personnel involved in goat husbandry. It is intended for use in areas where there is limited or no access to veterinary services. (Also see Sheep Health Handbook).

Winrock International Livestock Research and Training Center
Sheep Health Handbook

1983. 132 pages. 14 × 20 cm. Comb. HDC $4.65, LDC $4.65 plus airmail ($3.95) or surface mail ($1.00) postage.

This field handbook was designed to assist agricultural-development personnel involved in sheep husbandry. It is intended for use in areas where there is limited or no access to veterinary services. (Also see Goat Health Handbook.)

BOSTID

The Board on Science and Technology for International Development

Office of International Affairs
National Academy of Sciences/National Research Council

2101 Constitution Avenue, NW.
JH-210 Washington, D.C. 20418, USA
Cable: NARECO
Telex: 353001 BOSTID WSH
Telephone: 202/334-2639

What are the National Academcy of Sciences and the National Research Council?

The National Academy of Sciences (NAS) is a private, honorary society of scientists and engineers, dedicated to the furtherance of science and its uses for the general welfare. Although the Academy is not a federal agency, it is called upon by the terms of its 1983 charter to examine and report on any subject of science or technology upon request of any department of the federal government.

The National Research Council (NRC), serves as the operating arm of the National Academy of Sciences (NAS) and the National Academy of Engineering. The Institute of Medicine, a sister organization, also participates in the activities of the NRC.

The NRC is composed of eight major units, called assemblies and commissions. The Office of International Affairs has the broad function of conducting the international activities of the National Research Council.

What is the Board on Science and Technology for International Development (BOSTID)?

BOSTID is a division of the Office of International Affairs and is responsible for programs with developing countries. Established in 1969, BOSTID examines ways to apply science and technology to problems of economic and social development, through overseas programs, studies, advisory committees, and other mechanisms.

Participants in BOSTID activities work with counterpart groups in developing countries. This joint effort is directed toward strengthening local scientific and technological capabilities related to agriculture, environmental planning, energy, forestry, health, natural resource management and conservation, nutrition, water supply and quality, and other areas. Overseas activities also address the national organization and planning capabilities needed in applying science and technology to development. Studies examine specific development problems and suggest possible scientific and technological solution.

BOSTID's work relies on scientists and engineers who are selected for their expertise and who contribute their time and services as members of study panels and participants in overseas activities. BOSTID's permanent staff provides professional support and continuity and plans future programs.

What are the objectives of BOSTID's program?

- BOSTID seeks to help developing countries strengthen their own capabilities for dealing with important development-related problems and for moving toward greater scientific and technological self-reliance.
- BOSTID seeks to stimulate and support R&D within and among the developing countries on problems of high priority for development and human welfare.
- BOSTID seeks to provide developing countries with greater access to the scientific and technological know-how and expertise found in the United States and other countries.
- BOSTID seeks to provide a focal point within the U.S. scientific and technical community for assistance to the developing countries and to encourage greater cooperation between U.S. scientists and engineers and their colleagues in the Third World.

New price information for BOSTID publications

Title	Price ($)
Monographs	
More Water for Arid Lands: Promising Technologies and Research Opportunities	$6.50
Underexploited Tropical Plants with Promising Economic Value	$6.50
Making Aquatic Weeds Useful: Some Perspective for Developing Countries	$6.50
Postharvest Food Losses in Developing Countries	$7.50
Tropical Legumes: Resources for the Future	$8.00
Firewood Crops: Shrub and Tree Species for Energy Production	$7.00
Food, Fuel and Fertilizer from Organic Wastes	$6.50
Sowing Forests from the Air	$5.00
Winged Bean: A High Protein Crop for the Tropics (2nd edition)	$5.00
The Water Buffalo: New Prospects for an Underutilized Animal	$6.00
Mangium and Other Fast-Growing Acacias for the Humid Tropics	$5.00
Calliandra: A Versatile Small Tree for the Humid Tropics	$5.00
Crocodiles as a Resource for the Tropics	$5.00
Butterfly Farming in Papua New Guinea	$5.00
Little-Known Asian Animals with a Promising Economic Future	$6.00
Alcohol Fuels: Options for Developing Countries	$6.00
Producer Gas: Another Fuel for Motor Transport	$6.00
Firewood Crops: Shrub and Tree Species for Energy Production - Volume II	$5.50
Diffusion on Biomass Energy Technologies for Developing Countries	$6.00
Amaranth: Modern Prospects for an Ancient Crop	$5.50
Cauarinas: Nitrogen-Fixing Trees for Adverse Sites	$6.00
Jojoba: New Crop for Arid Lands	$6.00
Leucaena: Promising Forage and Tree Crop for the Tropics (2nd edition)	$6.00
Conferences, Workshop, and Symposia	
Proceedings, International Workshop on Survey Methodologies for Developing Countries	$7.00
Priorities in Biotechnology Research for International Development: Report of a Workshop	$7.00

GTZ

**Deutsche Gesellschaft für
Technische Zusammenarbeit (GTZ) GmbH**

Dag Hammarskjöld Weg 1, Postfach 5180, D-6236 Eschborn.
Cable: Germatec Eschborn Taunus
Telex: 417405 gtz d
Telefon: 06196-790

What is GTZ?

The Deutsche Gesellschaft für Technische Zusammenarbeit (GTZ) is a non-profit enterprise owned by the Federal Republic of Germany, represented by the Federal Minister for Economic Cooperation (BMZ) and the Federal Minister of Finance (BMF).

The GTZ has existed in its present form since January 1, 1975, when it took over the activities of two previous development agencies - the Bundesstelle für Entwicklungshilfe (BfE) and the Deutsche Förderungsgesellschaft für Entwicklungsländer (GAWI), to streamline operations.

The basis of the GTZ's work in official technical operation is a General Agreement concluded on December 12, 1974. This agreement clearly delineates the tasks between the Federal Ministry for Economic Cooperation (BMZ) which is responsible for development policy and the GTZ as implementing agency:
* The BMZ, the negotiating partner of the governments of the developing countries, defines the general development policy aims and the objectives of individual activities and supervises their implementation; it decides what projects are to be supported and sets the financial framework in its commissions to the GTZ.
* The GTZ carries out BMZ commissions under its own responsibility, either itself or by recalling in other state or private enterprises; it supports the BMZ in the further development of principles and instruments of development policy.

What is the purpose of GTZ?

GTZ puts development policy into practice. Most of its work is commissioned by the Government of the Federal Republic of Germany whose July 1980 "Policy Paper on German Cooperation with Developing Countries" defines technical cooperation as follows:
> "The purpose of technical cooperation is to enhance the performance of manpower and institutions in the developing countries. It involves the assignment of technical personnel and the supply of technical equipment. As a rule, technical cooperation are services granted on a non-repayable basis. In the more advanced developing countries, this cooperation is either partly or wholly against payment."

In addition to implementing Federal German development policy, the GTZ also takes direct commissions against payment from developing countries or from international organizations and it operates projects with its own funds on a limited scale.

GTZ is carrying out 2000 projects in more than 100 African, Asian and Latin American countries. More than 3000 staff members are involved in this work. The projects are implemented by GTZ's own field staff or by consultant firms and include the elaboration of studies and evaluations, handling of financial contributions, food aid etc. More than 10,000 supply consignments are made to projects every year. From 1975 to 1985 annual turnover rose from DM 460 to DM 1,015 million.

Main tasks

Technical cooperation projects span almost the entire economic and social spectrum of the developing countries. GTZ tasks and services are correspondingly broad:
- Assignment and funding of experts for agreed projects as well as advisers, instructors, specialists, appraisers and short-term experts;
- supply of equipment and materials for the facilities supported, supply of industrial and agricultural production inputs, provision of services and works;
- support of the reintegration in their home countries of expert manpower from developing countries who trained in the Federal Republic of Germany and wish to return home;
- salary topping up subsidies for German experts under direct contract to a developing country;
- coordination of the initial and further training of specialists and managerial personnel from developing countries either in their own countries, in other developing countries, in the Federal Republic of Germany or in other industrial countries;
- subsidization of the cost of training programmes carried out by companies in the developing countries;
- financial contributions to projects and programmes of efficient implementing agencies in developing countries.

To achieve these tasks the development projects are planned and implemented by three departments:

Department 1: Agriculture, Health and Rural Development
Department 2: Science and Technology, Education, Vocational Training, Industry and Trade
Department 3: Infrastructure

Agriculture, Health and Rural Development

The promotion and development of rural areas has always been a top priority in German development cooperation with developing countries. The activities of the Department of Agriculture, Health and Rural Development are therefore determined by the following goals:
- to raise and diversify production in agriculture, forestry and fisheries,
- to create more jobs by promoting the agricultural structure,

- to improve storage, distribution, marketing, and processing of agricultural products,
- to establish appropriate production, credit and marketing organizations,
- to intensify agricultural research and technology,
- to improve living conditions by supporting health, nutrition and family planning programmes.

The department cooperates closely with national and international research institutions and organizations in its research and development activities.

Publications

Experience gained in gtz projects is available to a wide circle of readers in gtz publications.

The GTZ series of publications can be obtained directly from the TZ-Verlagsgesellschaft mgH, Bruchwiesenweg 19, D-6101 Roßdorf 1
Telefon: Your code for Germany, then 6154-81119, against payment of the stipulated price, plus postage.

The following GTZ publications are no longer available:

Nr. 11, Nr. 40, Nr. 44, Nr. 53, Nr. 54, Nr. 67, Nr. 75, Nr. 90, Nr. 103, Nr. 112, Nr. 119, and Nr. 127.

Nr. 17
Lindau, Manfred

El Koudia/Marokko - Futterbau und Tierhaltung - Culture fourragère et entreien du bétail

1974. 74 Seiten. Deutsch und Französisch. 4 Abbildungen Bestell-Nr. 0017-Sr. DM 10,-. ISBN 3-9800030-7-8.

Die marokkanische Tierzuchtforschungsstation El Koudia, ein Vorhaben der Technischen Zussammenarbeit der Bundesrepublik Deutschland, das im Februar 1974 in die alleinige Verantwortung der Marokkaner übergegangen ist, wird in dieser Publikation vorgestellt. Die Schrift enthält Beiträge über die Situation der Futtererzeugung auf der Station El Koudia sowie Vorschläge zur Verbesserung besonders im Hinblick auf die Anlage von Dauerweiden. Es wird ferner auf Versuche zur Verbesserung von Matorral-Weideflächen durch unterschiedliche Bearbeitungsmaßnahmen und Unter-???

Nr. 72
Bruning, Dietrich C.

Population Planning in Pakistan - A Study of the Continuous Motivation System

1977. 300 pages. English. DM 18.50. ISBN 3-88085-143-7.

Rapid population growth has become a major concern of many Third World countries and thus also of Pakistan, where the annual population growth rate increased from 1.6% during the period from 1901-1911 to about 3% during the time from 1961-1972. To master the problem, the Pakistan government initiated a family planning programme in 1960; it was assisted by the Federal Republic of Germany. This publication describes the Pakistan family planning model.

Nr. 73
Seeber, G., Weidelt, H. J. and Banaag, V. S.

Dendrological Characters of Important Forest Trees from Eastern Mindanao

1979. 440 pages. English. DM 22.60. DM 20.50 ISBN 3-88085-068-2.

The largest continuous forest of the Philippines in eastern Mindanao is composed of a large number of variety of tree species. Many of these were unknown and could therefore not be identified reliably. This gap has now largely been filled.

This tree identification handbook owes its existence to a GTZ project in Davao City, Philippines. The book was to help in the elaboration on the Philippines of guidelines for the management of tropical forests since the need for well-planned forest utilization, even with regard to timber species not utilized to date, is increasingly being recognized.

The basic concept of the book was to summarize the dendrological characteristics permitting rapid and reliable identification in the field.

Nr. 93
Shaukat Ali, Chaudhary, Raman, Revri.

Weeds of North Yemen (Yemen Arab Republic)

1983. 412 pages. English. Bestel-Nr. 0093-Sr. DM 45,-. ISBN 3-88085-195-6.

Nr. 109
Nienhaus, Franz

Virus- and similar diseases in Tropical and Subtropical Areas

1981. 216 pages. English. DM 34.50. ISBN 3-88085-106-9

Nr. 129
Weidner, Herbert; Rack, Gisela.

Tables de détermination des principaux ravageurs des denrées entreposées dans les pays chauds (Tables to determine the principal pests of stored foods in hot countries).

198 pages. 123 ills. French. ISBN 3-88085-185-9. DM ___.

Nr. 146
Link, Rolf; Mouch, Moustapha.

Contributions a la biologie à la propagation et à la lutte contre les adventices au Maroc (Contributions to the biology, the spread and the control of weeds in Morocco).

1984. 160 pages. French. ISBN 3-88085-193-X. DM 20,50.

Nr. 147
Maydell, H. -J. von

Arbres et arbustes du Sahel - Leurs caractéristiques et leurs utilisations (Trees and shrubs of the Sahel - their characteristics and usages).

1983. 538 pages. Numerous colour ills. French. ISBN 3-88085-195-6. DM 97,50.

This book discusses about 120 different trees and shrubs which are native to the Sahel in order to point out the variety and usefulness of the natural vegetation, and at the same time warn against its overuse. Unless the ecological balance is reestablished the people living there and their animals will in the long run lose the basis on which their lives depend.

Nr. 155

Forschung und Entwicklung in der Technischen Zusammenarbeit mit Entwicklungsländern

1984. 272 pages. Deutsch. Bestell-Nr. 0155-Sr. DM 25,-. ISBN 3-88085-219-7.

Die vorliegende Zusammenstellung will eine Übersicht geben über die durch die GTZ durchgeführten Projekte mit Forschungs- und Entwicklungscharakter. Die in Kapitel 5 dieser Broschüre enthaltenen Übersichten geben einen Überblick über die Verteilung der dargestellten Projekte auf die Fachbereiche der GTZ und auf die einzelnen Länder. Die Verteilung auf die Fachbereiche der GTZ spiegelt sogleich eine sektorale Verteilung der Projekte wider.

Nr. 156

Entwicklung ohne Rückschläge Antworten eines Ökologen auf 20 Fragen im Hinblick auf die Ländiche Entwicklung in den Tropen und Subtropen

1984. 128 pages. Deutsch. 1 Farbfoto. 20 Strichzeichnungen Bestell-Nr. 0156-Sr. DM 16,-. ISBN 3-88085-228-6.

Um schwerwiegende Fehlplanungen bei ländlichen Entwicklungsvorhaben zu vermeiden, sollten ökologische, aber auch soziale, kulturelle, ökonomische und politische Bedingungen berücksichtigt werden. Der Göttinger Geobotaniker gibt aus seiner langjährigen Erfahrung praktische Lösungen für Probleme, die bei der Bewirtschaftung des Bodens in den Tropen und Subtropen auftreten.

Nr. 157
Whistler, W. Arthur.

Weed Handbook of Western Polynesia

1983. 152 pages. English. Zahlreiche Farbfotos Bestell-Nr. 0156-Sr. DM 41,-. ISBN 3-88085-227-8.

Nr. 158

Integrated Research Programme on Technologies for Rural Development

A German-Korean Cooperation Programme.
Englisch. 1984. 106 Seiten. 18 Farbfotos. 7 Strichzeichnungen. 13 Tabellen. Bestell-Nr. 0158-SR. DM 36,50. ISBN 3-88085-230-8.

Nr. 162
Busche, Detlef; Draga, Monika; Hagedorn, Horst.

Les sables éoliens - Modeles et Dynamique - La menace éolienne et son contrôle. Bibliographie annotée

1984. 800 pages. Französisch. Völlig überarbeitete und erweiterte Fassung der enlischen Erstausgabe (SR 65). Bestel-Nr. 0162-Sr. DM 32,-. ISBN 3-88085-242-1.

Nr. 167

Technical Cooperation in the Health Sector. Facts and Figures 1984.

In Vorbereitung. Englisch. 1984. ______ Seiten. Bestell-Nr. 0167-SR. DM. ______ ISBN 3-88085-253-7.

Nr. 168
Schlolaut, W. (Redaktion).

Abrégé de production cunicole tenant compte des conditions spécifiques des pays en développement

1985. 256 pages. Französisch. Übersetzung der SR 134. 28 Farbfotos. 4 SW-Fotos. 6 Abbildungen. 50 Tabellen. Bestell-Nr. 0168-Sr. DM 45,-. ISBN 3-88085-255-3.

Nr. 169
Schlolaut, W. (Redaktion).

Compendium for Rabbit Production - appropriate for conditions in developing countries

1985. 264 pages. English. Übersetzung der SR 134. 28 Farbfotos. 4 SW-Fotos. 6 Abbildungen. 50 Tabellen. Bestell-Nr. 0168-Sr. DM 30,-. ISBN 3-88085-256-1.

Nr. 177
Heber, G.; Teplitz, W.; Zeitinger, C.; Zieroth, G.

Biofuels for developing countries: promising strategy or dead end?

1985. 176 pages. English. Übersetzung der SR 153. Diagramme/Tabellen. Bestell-Nr. 0177-Sr. DM 18,-. ISBN 3-88085-261-8.

Nr. 178

Überleben trotz der Dürre. Statt Nahrungsmittelhilfe - Eigenproduktion

1985. 24 pages. Deutsch. 15 Farbfotos. 1 SW-Foto. 1 Karte. 3 Schaubilder. Bestell-Nr. 0178-Sr. DM 15,-. ISBN 3-88085-268-5.

Getreideimporte durch Nahrungsmittel-Eigenproduktion überflüssig zu machen, ist Hauptanliegen der mauretanischen ländlichen Entwicklungsgesellschaft SONADER. Die landesweit operierende "Société Nationale pour le Développement Rurale" verfolgt dabei zwei Ziele: die Verbesserung der Bewässerungslandwirtschaft im Senegaltal und die Beratung und Ansiedlung der maurischen Nomaden im Landesinnern. Die Broschüre schildert die Projekte und Maßnahmen, die im Rahmen der deutschen Technischen Zusammenarbeit unterstützt wurden.

Nr. 179
Steiner, Kurt G.

Cultures associées dans les petites exploitations agricoles tropicales en particulier en Afrique de l'Ouest

1985. 348 pages. Französisch. Übersetzung der SR 137. 59 Tabellen/ Übersichten. 54 Strichabbildungen. Bestell-Nr. 0179-Sr. DM 35,-. ISBN 3-88085-271-5.

INTSOY

International Soybean Program
Programme International pour le Soja
Programa Internacional de la Soya

University of Illinois at Urbana-Champaign
113 Mumford Hall
1301 West Gregory Drive
Urbana, Illinois 61801, USA
Telex: 206957
Cable: INTSOY
Telephone: (217) 333-6422

Introduction and Objectives

INTSOY seeks to improve human nutrition around the world through the use of soybeans, a legume rich in protein and calories. Cooperating with likeminded regional, national, and international organizations, we work toward this goal through research, cultivar testing and breeding programs, feasibility studies, publications, regional and international conferences, training courses, and study programs.

INTSOY provides timely information about cultivar adaptation, crop management and protection, seed storage, and the processing and use of soy food. We offer direct access to a worldwide collection of soybean germplasm, and conduct a breeding program under varied ecological conditions. Through training programs, conferences, and publications, we are extending practical and timely information to soybean worksers.

The United States Agency for International Development (USAID) has provided the core funding since INTSOY's creation in April 1973. Special grants from FAO, UNDP,M UNICEF, CARE, and IBPGR have supported a range of special projects.

INTSOY, el Programa Internacional de la Soya, está tratando de mejorar la nutrición humana en todo el mundo a través del uso de la soya, una léguminosa rica en protéina y calorías. Un programa de la Universidad de Illinois en Urbana-Champaign, INTSOY coopera con organizaciones nacionales, regionales e internacionales para la expansión del uso de la soya. Comuníquese con nosotros si desea mayor información sobre los programas de INTSOY.

INTSOY, un Programme International pour le Soja, cherche à améliorer l'alimentation humaine dans le monde par 'lusage du soja, un légume riche en protéines et en calories. INTSOY est un programme de l'Université d'Illinois en cooperátion avec des organisations nationales, régionales et internationales pour la généralisation de 'lutilisation du soja. N'hésitez pas à nous contacter pour tous renseignements complementaires concernant les programmes d'INTSOY.

Miscellaneous Reports and Brochures

J. A. Jackobs, C. A. Smyth and D. R. Erickson

International Soybean Variety Experiment, Eleventh Report of Results, 1984

1986. ____ pages. 21.5 × 28 cm. Paperback. English (INTSOY Series No. 29). In preparation.

ICIMOD

International Centre for Integrated Mountain Development

P.O. Box 3226, Kathmandu, Nepal
Cable: ICIMOD Kathmandu
Telex: NP 2245 SATA
Telephone: 522819, 521575

General Information

The primary objective of the Centre is to promote economically and environmentally sound development in the Hindu Kush-Himalayas and to improve the well-being of the local population. This region includes, partially or totally, *Afghanistan, Bangladesh, Bhutan, Burma, China, India, Nepal* and *Pakistan.*

ICIMOD is a centre for multi-disciplinary documentation, training, and applied research as well as for consultative services on resource management and development activities in mountain regions.

The staff comprises a multi-national team of scientists, scholars, and experts in the many fields of mountain development. Appointed primarily from the countries of the Hindu Kush-Himalaya region, the long-term staff work in cooperation with colleagues from around the world recruited for specific short-term contribution.

The establishment of the Centre is based on an agreement between His Majesty's Government of Nepal and the United Nations Educational Scientific and Cultural Organization (UNESCO) signed in September 1981. At present the Centre is being sponsored by the following four parties: His Majesty's Government of Nepal, the Government of the Federal Republic of Germany, the Government of Switzerland, and UNESCO.

The Centre is located in Kathmandu, the capital of the Kingdom of Nepal, and enjoys the status of an autonomous international organization.

Monographs

International Centre for Integrated Mountain Development

Occasional Papers

Number of pages varies. 21 X 30 cm. Soft cover. $5.00. Air postage cost is $1.50. Surface mail is free.

The objective of the Occasional Paper series is to promote the exchange of practical field experience having major policy implications for the Hindu Kush-Himalaya region.

No.1

Brian Carson

Erosion and Sedimentation Processes in the Nepalese Himalaya

August 1985. 39 pages.

The relations between human activities and erosion and sedimentation processes in the Nepalese Himalaya are analyzed.

No. 2

B.B. Pradhan

Integrated Rural Development Projects in Nepal — A Review

December 1985. 58 pages.

This synthesis of integrated rural development projects in Nepal draws conclusions regarding performance and areas for improvement.

Proceedings

Jeffrey A. McNeely, James W. Thorsell and Suresh R. Chalise, editors

People and Protected Areas in the Hindu Kush-Himalaya

December 1985. 112 pages. 21 X 29 cm. Soft cover.

The International Workshop on the Management of National Parks and Protected Areas in the Hindu Kush-Himalaya was held 6-11 May, 1985, in Kathmandu, Nepal. Convened by the International Centre for Integrated Mountain Development in partnership with the King Mahendra Trust for Nature Conservation, the Workshop explored the question of how to involve local people more effectively in mountain conservation. The proceedings include: Opening Addresses, Defining the Problems and Suggesting Solutions, Case Studies, Protected Area Management Issues, and International Perspectives.

Kk Pandey, Convenor

International Workshop on Watershed Management in the Hindu Kush-Himalaya Region: A Report

March 1986. 50 pages. 17 X 25 cm. Soft cover.

The International Workshop on Watershed Management in the Hindu Kush-Himalaya Region, held in Chengdu, China, 14-19 October 1985, was the first international workshop to concentrate specifically on mountain watersheds and mountain communities of the region.

Jointly organised by ICIMOD and the Commission for Integrated Survey of Natural Resources of the Chinese Academy of Sciences, the workshop reviewed and evaluated progress made in the field of watershed management in the region. The proceedings include: Introduction, Overview, Discussion Papers, and Paper Summaries.

Miscellaneous

ICIMOD

**About ICIMOD
Mountain Development 2000: Challenges and Opportunities, Proceedings of the First International Symposium and the Inauguration of ICIMOD, 15 December 1983.**

1984. 123 pages. 21 × 28.5 cm. Soft cover.

The proceedings deal with the environmental degradation and overuse of natural resources in the Hindu Kush-Himalaya. The authors, mainly scientists, practitioners, and decision-makers, discuss barriers and ways to an ecologically and economically sound development of the mountain areas. Special attention is given to the work programme of ICIMOD.

Periodicals

ICIMOD Newsletter

8 pages. 21 × 29.5 cm.

Published triannually and distributed free of charge.

IIMI

International Irrigation Management Institute

Headquarters:
IIMI, Digana Village via Kandy, Sri Lanka
Telex: 22318 IIMIHQ CE
Telephone: (08) 74274

Liaison Office:
IIMI, P.O. Box 2075, 5A Schofield Place, Colombo 5, Sri Lanka
Telephone: (01) 589933

International Irrigation Management Institute

The International Irrigation Management Institute (IIMI) is an autonomous, non-profit international organization chartered in Sri Lanka in 1984 to conduct research, provide opportunities for professional development, and communicate information about irrigation management. Through collaboration, IIMI seeks ways to strengthen independent national capacity to improve the management and performance of irrigation systems in developing countries.

IIMI's research program aims at deriving methodologies and conceptual understandings that result in better management of irrigation resources. Researchers seek to understand the ways in which local conditions - physical and social -affect the performance of irrigation systems. Attention is focused on the whole system through a multi-disciplinary approach.

IIMI's training program is designed to strengthen leadership and management skills among professionals responsible for planning and managing irrigation systems. This is accomplished through workshops and conferences, and support for graduate students and post-doctoral fellowships in innovative irrigation management.

IIMI's information program supports IIMI researchers and an international network of people interested in irrigation management. The Communication and Publication Office produces publications on irrigation management topics, provides documentation services through IIMI's computerized database, and maintains the Institute's library.

IIMI's headquarters is in Digana Village near Kandy, about 130 km east of Colombo and central to some of Sri Lanka's extensive irrigation projects.

For further information, please write: IIMI, Digana Village via Kandy, Sri Lanka. Telephone (08) 74274, 74253, and 74334. Telex 22318 IIMIHQ CE.

Annual Reports

International Irrigation Management Institute
IIMI Annual Report for 1984-1985

1986. Digana Village, Sri Lanka: IIMI pub.

Conferences, Symposia, and Workshop Proceedings

International Irrigation Management Institute and Joint WHO/FAO/
UNEP Panel of Experts on Environmental Management for Vector
Control (PEEM)

**Proceedings of the Workshop on Irrigation and Vector
Borne Disease Transmission, October 13-17 1985**

1986. Digana Village, Sri Lanka: IIMI pub.

International Irrigation Management Institute and Ministry of Lands
and Land Development of the Government of Sri Lanka

**Proceedings of a Workshop on Participatory Management
in Sri Lanka's Irrigation Schemes, May 15-17 1986**

1986. Digana Village, Sri Lanka: IIMI pub.

International Irrigation Management Institute and Water and Energy
Commission Secretariat (WECS) of the Ministry of Water Resources,
Government of Nepal

**Proceedings of a Workshop on Public Intervention in
Farmer-managed Irrigation Systems, July 4-6 1986**

1986. Kathmandu, Nepal. Digana Village, Sri Lanka: IIMI pub.

Periodicals/Serials

Research Paper Series

No. 1
Small, Leslie E. and R. Barker. 1985. Summary of a Joint IIMI/Water
Management Synthesis II workshop on research priorities for irrigation
management in Asia, Digana Village, Sri Lanka, January 6-11, 1985.
(English and French; French edition in press).

No. 2
Groenfeldt, D. (ed.). 1986. Proceedings from a workshop on selected
irrigation management issues, Digana Village, Sri Lanka, July 15-19,
1985.

No. 3
Wolf, J. and D. Merrey. 1986. Irrigation management in Pakistan: Four
papers. Digana Village, Sri Lanka: IIMI pub.

No. 4
Chambers, Robert and Ian Carruthers. 1986. Rapid appraisal to
improve canal irrigation performance: Experience and options. Digana
Village, Sri Lanka: IIMI pub.

Management Briefs

No. 1
Rao, P.S. and A. Sundar.

Managing Main System Water Distribution

June, 1966, Digana Village, Sri Lanka.

No. 2
Chambers, Robert and Ian Carruthers.

Rapid rural appraisal for irrigation system

July, 1986, Digana Village, Sri Lanka.

Periodicals

International Irrigation Management Institute.

Letter from the Director General

1986. Vol. 1.

Overseas Development Institute-IIMI.

Irrigation Management Network Newsletter

1986. Vol. 86.

Miscellaneous

International Irrigation Management Institute

IIMI: A Descriptive Brochure

November 1984. English and French; (French edition in press).

Appendix A

Distributing bodies for Unesco coupons

In many countries, the lack of foreign currency hinders the importation of publications. If your country is included in the following list, you can buy Unesco coupons in your local currency to pay for the publications of IARCs. The coupons are issued in values of 1, 10, 30, 100, and 1000 US Dollars and available at the official United Nations rate of exchange on the day of purchase. The distributors may charge an additional amount for administrative costs, which may not exceed 5% of the value of the coupons.

Unesco coupons are a means of payment. Therefore take all precautions to avoid loss. In case of loss, communicate the serial numbers of the lost coupons to:

Bureau des Bons de l'Unesco, 7 Place de Fontenoy, 75700 Paris, France.

The Unesco replaces them if they have not been redempted within the following six months.

Organismos distribuidores de bonos de la Unesco

En muchos países la falta de divisas dificulta la importación de publicaciones. Si su país está incluido en la lista siguiente, usted puede comprar los cupones de la Unesco en su moneda local para pagar las publicaciones. Los cupones son emitidos en valores de 1, 10, 30, 100, y 1000 dólares estadounidenses y disponibles al cambio oficial de las Naciones Unidas en el día de la compra. Los distribuidores pueden cobrar una suma adicional para los gastos administrativos, que no deberá exceder de 5% del valor de los cupones.

Los Cupones de la Unesco son un medio de pago. Por ello tome todas las precauciones para evitar su pérdida. Si ésta ocurre, comunique los números de serie de los cupones perdidos a:

Bureau des Bons de l'Unesco, 7 Place de Fontenoy, 75700 Paris, France.

La Unesco los reemplazara si no son redimidos en los siguientes seis meses.

Organismes distributeurs des bons de l'Unesco

Dans plusieurs pays, le manque de devises empêche l'importation de publications. Si votre pays est inclus dans la liste suivante, vous pouvez acheter des bons de l'Unesco dans votre monnaie locale pour payer les publications. Les bons sont émis pour les valeurs de 1, 10, 30, 100, et 1000 US Dollars, et disponibles au taux de change officiel des Nations Unies le jour de l'achat. Les distributeurs peuvent réclamer un supplément pour frais administratifs, qui ne doit pas excéder 5% de la valeur des bons.

Les bons de l'Unesco sont un moyen de paiement. Donc, prenez vos précautions afin de ne pas les perdre. En cas de perte, communiquez les numéros de série des bons égarés à:

Bureau des Bons de l'Unesco, 7 Place de Fontenoy, 75700 Paris, France.

L'Unesco en émettra de nouveaux, s'ils n'ont pas été présentés dans les six mois suivant le perte.

Algeria/Algérie/Argelia
Commission nationale algérienne pour
l'Unesco et l'Alesco
Ministère de l'enseignement supérieur
et de la recherche scientifique.
23, avenue de la Robertsau
TELEMLY-ALGER

Angola
Commission nationale angolaise pour l'Unesco
Ministère de l'Education
C.P. 1281
LUANDA

Argentina/Argentine
Comisión Nacional Argentina para la Unesco
Paseo Colón 533, 7° piso
1063 BUENOS AIRES

Austria/Autriche
Hauptverband des Ö₃sterreichischen Buchhandels
Grünangergasse 4,
A 1010 WIEN 1

Bangladesh
Bangladesh National Commission for Unesco
Ministry of Education
House No. 60 - Road No. 2A. Dhanmondi
DHAKA 5

Brazil/Brésil/Brasil
Instituto Brasileiro de Educação, Ciencia e Cultura
(I.B.E.C.C.), Praia de Botafogo, 186 Terreo,
Sals 101/102
RIO DE JANEIRO

Bulgaria/Bulgarie
Commission nationale de la République populaire
de Bulgarie pour l'Unesco
Boîte postale 386
SOFIA

Burma/Birmanie/Birmania
Burmese National Commission for Unesco
Dept. of Higher Education
Ministry of Education, Office of the Ministers,
Theinbyu Street,
RANGOON

Burundi
Commission nationale du Burundi pour l'Unesco
Ministère de l'éducation nationale et de la culture,
Boîte postale 1900
BUJUMBURA

Chili/Chile
Conicyt
Canadá 308 - Casilla 297, V
SANTIAGO DE CHILE 9

Colombia/Colombie
I.C.E.T.E.X.
Oficina de Relaciones Nacionales o Internacionales
Apartado Aereo 5735
BOGOTA D.E.

Congo
Commission nationale congolaise pour l'Unesco
Boîte postale 493
BRAZZA VILLE

Cuba
Comisión Nacional Cubana de la Unesco
Avenida Kohly 151, Esq. 32, Nuevo Vedado
LA HABANA

Egypt/Egypte/Egipto
Unesco Representative in the Arab Republic
of Egypt. 8. Abdel Rahman Fahmy Street,
Garden City
CAIRO

Ethiopia/Ethiopie/Etiopia
Ethiopian National Agency for Unesco
P.O. Box 2996
ADDIS ABABA

France/Francia
(y compris les départements français d'Outre mer)
Foyer international d'accueil de Paris
30, rue Cabanis
75014 Paris

et Centre national de la recherche scientifique
Service des Bons de l'Unesco
23, rue de Maroc
75940 PARIS CEDEX 19

Gambia/Gambie
Gambia National Commission for Unesco
Ministry of Education, Youth, Sports and Culture
Bedford Place Building
BANJUL

Ghana
Ghana National Commission for Unesco
Ministry of Education, P.O. Box 2739
ACCRA

Guinea/Guinée
Commission nationale guinéenne pour l'Unesco
Ministère de l'enseignement supérieur
et de la recherche scientifique
Boîte postale 964
CONAKRY

Guinea-Bissau/Guinée-Bissau
Commission nationale de Guinée-Bissau pour l'Unesco
Ministère de l'Éducation, de la Culture et des Sports
C.P. 353
BISSAU

Guyana/Guyane
United Nations Development Programme in Guyana
P.O. Box 10960
GEORGETOWN

India/Inde
Indian National Commission for Co-operation with
Unesco, Ministry of Education and Social Welfare
"C" Wing, Shastri Bhawan
NEW DELHI 110001

Islamic Republic of Iran/Rép. islamique d'Iran
Iranian National Commission for Unesco
1188 Enghelab avenue
Rostam Give Building
P.O. Box 11365-4498 - Zip code 13158
TEHERAN

Iraq/Irak
Iraq National Commission for Unesco
Ministry of Education
BAGHDAD

Ivory Coast/Côte d'ivoire/Costa de Marfil
Commission nationale ivoirienne pour l'Unesco
Ministère de l'éducation nationale et
de la recherche scientifique
15, avenue Noguès
01, B.P. 297
ABIDJAN 01

Jamaica/Jamaïque
Jamaica National Commission for Unesco
30 Grenade Crescent
KINGSTON 5

Japan/Japon/Japon
Japan Society for the Promotion of Science
Yamato Building
5-3-1 Kojimachi Chiyoda-ku
TOKYO 102

Kenya
Kenya National Commission for Unesco
Ministry of Education, Science and Technology
P.O. Box 30040
NAIROBI

Republic of Korea/République de Corée/República de Corea
Korea Exchange Bank
I.P.O. 2924
SEOUL

Madagascar
Commission nationale malgache pour l'Unesco
Ministère de l'enseignement secondaire et de
l'éducation de base
11, Naka Rabemanantsoa - Behoririka
Boîte postale 331
TANANARIVE - ANTANANARIVO (101)

Malawi
Malawi National Commission for Unesco
Taurus House - P.O. Box 30278, Capital City
LILONGWE 3

Mauritania/Mauritanie
Commission nationale de la République islamique
de Mauritanie pour l'Unesco
B.P. 196
NOUAKCHOTT

Monaco/Mónaco
Commission nationale pour l'Éducation
la Science et al Culture
Ministère d'État
MONACO-Ville

Mongolia/Mongolie
Commission nationale de la Mongolie pour l'Unesco
Ministère des Affaires étrangères
OULAN BATOR

Morocco/Maroc/Marruecos
Commission nationale marocaine pour l'éducation
la science et al culture
Boîte postale 420 - 19 rue Oqba
Agdal - RABAT

et OMAFAY'S
16, rue Tanta
RABAT

Mozambique
National Commission for Unesco of the People's Republic of
Mozambique
Ministry of Education and Culture
P.O. Box 3674
MAPUTO

Nepal/Népal
Nepalese National Commission for Unesco
Ministry of Education, Kaiser Mahal, Kantipath
KATHMANDU

Nicaragua
United Nations Development Programme
Apartado postal 3260
MANAGUA

Nigeria/Nigéria
Nigerian National Commission for Unesco
Federal Ministry of Education
Science and Technology
Victoria Island
LAGOS

Pakistan/Pakistán
Pakistan National Commission for Unesco
Ministry of Education
ISLAMABAD

Peru/Pérou
No distributing body in Peru

Philippines
United Nations Development Programme in the Philippines
P.O. Box 1864
MANILA

and 7285 ADC, MIA Rd. Pasay City
METRO MANILA

Poland/Pologne
Commission nationale polonaise pour l'Unesco
Palac Kultury i Nauki (17 pietro)
00-901 VARSOVIE

Portugal
Comisâo Nacional para a Unesco
(Commission nationale portugaise pour l'Unesco)
Ministère des Affaires étrangerès
Avenida Infante Santo N. ° 42-5°
1300 LISBONNE

Romania/Roumanie/Rumania
Commission nationale de la République socialiste de Roumanie pour
l'Unesco
Soseaua Kisaleff N° 47 - B.P. 71268
BUCAREST

Rwanda
Commission nationale rwandaise pour l'Unesco
Ministère de l'enseignement supérieur
et de la Recherche scientifique
B.P. 1326
KIGALI

Sierra Leone
Sierra Leone National Commission for Unesco
Ministry of Education
FREETOWN

Spain/Espagne/España
Instituto Nacional del Libro Español
Calle Santiago Rusiñol, 8
MADRID 3

Sri Lanka
Sri Lanka National Commission for Unesco
Ministry of Justice
238 Hulftsdorp Road
COLOMBO 12

Sudan/Soudan/Sudán
United Nations Development Programme in the Sudan
P.O. Box 913
KHARTOUM

Syrian Arab Republic/République arabe syrienne/República Arabe
Siria
Centre d'etudes et de recherches scientifiques
Boîte postale 4470
DAMAS

Tanzania/Tanzanie
United Republic of Tanzania Unesco National Commission
Ministry of National Education
P.O. Box 20384
DAR ES SALAAM

Togo
Commission nationale de la République togolaise pour l'Unesco
Ministère de l'Éducation nationale et de la
Recherche scientifique
LOMÉ

Tunisia/Tunisie/Túnez
Commission nationale tunisienne pour l'Éducation, la Science et al
Culture
Ministère de l'éducation nationale
22, rue d'Angleterre 1280
1055 TUNIS R.P.

Uganda/Ouganda
Ministry of Finance Planning and Economic Development
P.O. Box 7086
KAMPALA

United Kingdom/Royume-Uni/Reino Unido
(including British Commonwealth and Trust Territories)
Lloyds Bank International Ltd.
P.O. Box 241 - 100 Pall Mall
LONDON SW1 5HP

Uruguay
C.O.N.I.C.Y.T., Sarandy 444, p.4,
Cascilla de correo 1869
MONTEVIDEO

Yugoslavia/Yougoslavie
Commission yougoslave pour la coopération avec l'Unesco
Mose Pijade 8 (6éme étage)
BELGRADE

Zaire/Zaïre
Commission nationale zaïroise pour l'Unesco
Commissariat de'État a l'enseignement primaire et secondaire, Boîte
postale 3163
KINSHASA - Gombe

Zambia/Zambie
Zambia National Commission for Unesco
Ministry of Education and Culture
P.O. Box 50619
LUSAKA

Zimbabwe
Unesco Desk, Ministry of Education and Culture
P.O. Box 8022
CAUSEWAY

Addendum

Publications of the International Agricultural Research and Development Centers

1985. 691 pages. 13 × 21 cm. Paperback. All nations, US$10.20 (includes airmail postage and handling); Philippines, ₱80.35 (includes surface mail and handling).

In 1985 the International Rice Research Institute published *Publications of the International Agricultural Research and Development Centers,* a 691-page catalog of titles published by 13 Centers supported by the Consultative Group on International Agricultural Research (CGIAR); 8 other IARCs; the Board of Science and Technology for International Development (BOSTID) of the U.S. National Academy of Sciences; and the German Agency for Technical Cooperation (GTZ). The catalog includes 162 pages of in-depth index to help reader locate all publications in certain fields (i.e. cytogenetics, insects resistance, maize).

Supplement to Publications of the International Agricultural Research Development Centers

1986. 167 pages. 13 × 21 cm. Perfect binding. US$4.00 (includes airmail postage and handling); Philippines, ₱40.00 (includes surface mail and handling)

The 1986 Supplement to Publications of the International Agricultural Research Development Centers includes only new titles that are not in the larger 1985 catalog. The two catalogs are the only compilations of the major Center publications and, collectively, are probably the largest catalog of titles on Third World agriculture in existence.